Icons
of Evolution

About the Author

Jonathan Wells is no stranger to controversy. After spending two years in the U.S. Army from 1964 to 1966, he entered the University of California at Berkeley to become a science teacher. When the Army called him back from reserve status in 1968, he chose to go to prison rather than continue to serve during the Vietnam War. He subsequently earned a Ph.D. in religious studies at Yale University, where he wrote a book about the nineteenth-century Darwinian controversies. In 1989 he returned to Berkeley to earn a second Ph.D., this time in molecular and cell biology. He is now a senior fellow at Discovery Institute's Center for the Renewal of Science and Culture (www.discovery.org/crsc) in Seattle, where he lives with his wife, two children, and mother. He still hopes to become a science teacher.

Icons
of Evolution

Science or Myth?

Why Much of What
We Teach About Evolution
Is Wrong

JONATHAN WELLS

ILLUSTRATED BY
JODY F. SJOGREN

Since 1947
REGNERY
PUBLISHING, INC.
An Eagle Publishing Company • Washington, DC

First paperback edition 2002

Library of Congress Cataloging-in-Publication Data

    Wells, Jonathan.
        Icons of evolution: science or myth?: why much of what we teach about evolution is wrong/by Jonathan Wells.
        p. cm.
        ISBN 0-89526-200-2
        1. Evolution (Biology)   I. Title

    QH366.2.W45 2000
    576.8—dc21   00-062544

Published in the United States by
Regnery Publishing, Inc.
An Eagle Publishing Company
One Massachusetts Avenue, NW
Washington, DC 20001

Visit us at www.regnery.com

Distributed to the trade by
National Book Network
4720-A Boston Way
Lanham, MD 20706

Printed on acid-free paper
Manufactured in the United States of America

10 9 8 7 6

Books are available in quantity for promotional or premium use. Write to Director of Special Sales, Regnery Publishing, Inc., One Massachusetts Avenue, NW, Washington, DC 20001, for information on discounts and terms or call (202) 216-0600.

For Josie and Peter

The iconography of persuasion strikes even closer than words to the core of our being. Every demagogue, every humorist, every advertising executive, has known and exploited the evocative power of a well-chosen picture....
But many of our pictures are incarnations of concepts masquerading as neutral descriptions of nature. These are the most potent sources of conformity, since ideas passing as descriptions lead us to equate the tentative with the unambiguously factual.

—Stephen Jay Gould, *Wonderful Life*
(New York: W. W. Norton, 1989, p. 28)

Contents

Preface

During my years as a physical science undergraduate and biology graduate student at the University of California, Berkeley, I believed almost everything I read in my textbooks. I knew that the books contained a few misprints and minor factual errors, and I was skeptical of philosophical claims that went beyond the evidence, but I thought that most of what I was being taught was substantially true.

As I was finishing my Ph.D. in cell and developmental biology, however, I noticed that all of my textbooks dealing with evolutionary biology contained a blatant misrepresentation: Drawings of vertebrate embryos showing similarities that were supposed to be evidence for descent from a common ancestor. But as an embryologist I knew the drawings were false. Not only did they distort the embryos they purported to show, but they also omitted earlier stages in which the embryos look very different from each other.

My assessment of the embryo drawings was confirmed in 1997, when British embryologist Michael Richardson and his colleagues published an article in the journal *Anatomy and Embryology,* comparing the textbook drawings with actual embryos. Richardson was subsequently quoted in the leading American journal *Science* as saying: "It looks like it's turning out to be one of the most famous fakes in biology."

Yet most people remain unaware of the truth, and even biology textbooks published after 1997 continue to carry the faked drawings. Since then, I have discovered that many other textbook illustrations distort the evidence for evolution, too. At first, I found this hard to believe. How could so many textbooks contain so many misrepresentations for so long? Why hadn't they been noticed before? Then I discovered that other biologists *have* noticed most of them, and have even criticized them in print. But their criticisms have been ignored.

The pattern is consistent, and suggests more than simple error. At the very least, it suggests that Darwinism encourages distortions of the truth. How many of these distortions are unconscious and how many are deliberate remains to be seen. But the result is clear: Students and the public are being systematically misinformed about the evidence for evolution.

This book is about that evidence. To document it, I quote from the peer-reviewed work of hundreds of scientists, most of whom believe in Darwinian evolution. When I quote them, it is not because I want to make it sound as though they reject Darwin's theory; most of them do not. I quote them because they are experts on the evidence.

Wherever possible, I have avoided technical language. For those who want more details, I include extensive notes at the end of the book referring to the scientific literature. The notes are not intended to be exhaustive (except where they list sources of quotations), but to aid readers who want to pursue matters further.

The chapters are followed by two appendices. The first critically evaluates ten widely-used biology textbooks, from the high school to the graduate level. The second suggests warning labels, like those used on packs of cigarettes, that schools might want to place in their teaching materials to alert students to the misrepresentations.

Many people were kind enough to review and comment on the manuscript. Those who assisted me with technical details in the indicated sections or chapters include: Lydia McGrew (Introduction); Dean Kenyon and Royal Truman (The Miller-Urey Experiment); John Wiester (the Cambrian explosion, in The Tree of Life); W. Ford Doolittle (molecular phylogeny, in The Tree of Life); Brian K. Hall (Homology); Ashby Camp and Alan Feduccia (*Archaeopteryx*); Theodore D. Sargent (Peppered Moths); Tony Jelsma (Darwin's Finches); Edward B. Lewis (genetics of triple mutants, in Four-winged Fruit Flies); and James Graham (human origins, in The Ultimate Icon). Listing these people here does not imply that they endorse my views. On the contrary, many of them will disagree with my conclusions and recommendations. But for these fine people, science is the search for truth, and I am indebted to them for helping me get the facts straight. Of course, any errors that remain are my fault, not theirs.

People who patiently read and commented on major portions of the manuscript include (in alphabetical order) Tom Bethell, Roberta T. Bidinger, Bruce Chapman, William A. Dembski, David K. DeWolf, Mark Hartwig, Phillip E. Johnson, Paul A. Nelson, Martin Poenie, Jay Wesley Richards, Erica Rogers, Jody F. Sjogren (who also did most of the illustrations), Lucy P. Wells, and John G. West, Jr. Some of these readers helped me with scientific content, but all of them helped me to make the book readable. If there are still errors or rough spots, it is because I failed to follow all of their excellent advice.

I am grateful for research assistance from many people, especially Winslow G. Gerrish and William Kvasnikoff, and from staff members of the Natural Sciences and Health Sciences Libraries at the University of Washington, Seattle. Research funding for the book was generously provided by the Center for the Renewal of

Science and Culture (www.crsc.org), a project of the Discovery Institute in Seattle.

In addition to the people named above, other scientists at universities in the United States, Canada, and the United Kingdom helped with various parts of the manuscript, but prefer to remain anonymous. In several cases, they chose anonymity because their careers might suffer at the hands of people who strongly disagree with the conclusions of this book. For those scientists, public acknowledgment will have to wait.

Seattle, Washington
July 2000

CHAPTER I

Introduction

"Science is the search for the truth," wrote chemist Linus Pauling, winner of two Nobel prizes. Bruce Alberts, current president of the U. S. National Academy of Sciences, agrees. "Science and lies cannot coexist," said Alberts in May 2000, quoting Israeli statesman Shimon Peres. "You don't have a scientific lie, and you cannot lie scientifically. Science is basically the search of truth."

For most people, the opposite of science is myth. A myth is a story that may fulfill a subjective need, or reveal something profound about the human psyche, but as commonly used it is not an account of objective reality. "Most scientists wince," writes former *Science* editor Roger Lewin, "when the word 'myth' is attached to what they see as a pursuit of the truth." Of course, science has mythical elements, because all human enterprises do. But scientists are right to wince when their pronouncements are called myths, because their goal as scientists is to minimize subjective storytelling and maximize objective truth.

Truth-seeking is not only noble, but also enormously useful. By providing us with the closest thing we have to a true understanding of the natural world, science enables us to live safer, healthier and more productive lives. If science weren't the search for truth, our bridges wouldn't support the weight we put on them, our lives wouldn't be as long as they are, and modern technological civilization wouldn't exist.

Storytelling is a valuable enterprise, too. Without stories, we would have no culture. But we do not call on storytellers to build bridges or perform surgery. For such tasks, we prefer people who have disciplined themselves to understand the realities of steel or flesh.

The discipline of science

How do scientists discipline themselves to understand the natural world? Philosophers of science have answered this question in a variety of ways, but one thing is clear: Any theory that purports to be scientific must somehow, at some point, be compared with observations or experiments. According to a 1998 booklet on science teaching issued by the National Academy of Sciences, "it is the nature of science to test and retest explanations against the natural world."

Theories that survive repeated testing may be tentatively regarded as true statements about the world. But if there is persistent conflict between theory and evidence, the former must yield to the latter. As seventeenth-century philosopher of science Francis Bacon put it, we must obey Nature in order to command her. When science fails to obey nature, bridges collapse and patients die on the operating table.

Testing theories against the evidence never ends. The National Academy's booklet correctly states that "all scientific knowledge is, in principle, subject to change as new evidence becomes available." It doesn't matter how long a theory has been held, or how many scientists currently believe it. If contradictory evidence turns up, the theory must be reevaluated or even abandoned. Otherwise it is not science, but myth.

To ensure that theories are tested objectively and do not become subjective myths, the testing must be public rather than private. "This process of public scrutiny," according to the National Academy's booklet, "is an essential part of science. It works to eliminate individual bias and subjectivity, because others must also be able to determine whether a proposed explanation is consistent with the available evidence."

Within the scientific community, this process is called "peer review." Some scientific claims are so narrowly technical that they can be properly evaluated only by specialists. In such cases, the "peers" are a handful of experts. In a surprising number of instances, however, the average person is probably as competent to make a judgment as the most highly trained scientist. If a theory of gravity predicts that heavy objects will fall upwards, it doesn't take an astrophysicist to see that the theory is wrong. And if a picture of an embryo doesn't look like the real thing, it doesn't take an embryologist to see that the picture is false.

So an average person with access to the evidence should be able to understand and evaluate many scientific claims. The National Academy's booklet acknowledged this by opening with Thomas Jefferson's call for "the diffusion of knowledge among the people. No other sure foundation can be devised for the preservation of freedom and happiness." The booklet continued: "Jefferson saw clearly what has become increasingly evident since then: the fortunes of a nation rest on the ability of its citizens to understand and use information about the world around them."

U. S. District Judge James Graham affirmed this Jeffersonian wisdom in an Ohio newspaper column in May 2000. Graham wrote: "Science is not an inscrutable priesthood. Any person of reasonable intelligence should, with some diligence, be able to understand and critically evaluate a scientific theory."

Both the National Academy's booklet and Judge Graham's newspaper column were written in the context of the present controversy over evolution. But the former was written to defend Darwin's theory, while the latter was written to defend some of its critics. In other words, defenders as well as critics of Darwinian evolution are appealing to the intelligence and wisdom of the American people to resolve the controversy.

This book was written in the conviction that scientific theories in general, and Darwinian evolution in particular, can be evaluated by any intelligent person with access to the evidence. But before looking at the evidence for evolution, we must know what evolution is.

What is evolution?

Biological evolution is the theory that all living things are modified descendants of a common ancestor that lived in the distant past. It claims that you and I are descendants of ape-like ancestors, and that they in turn came from still more primitive animals.

This is the primary meaning of "evolution" among biologists. "Biological evolution," according to the National Academy's booklet, "explains that living things share common ancestors. Over time, evolutionary change gives rise to new species. Darwin called this process 'descent with modification,' and it remains a good definition of biological evolution today."

For Charles Darwin, descent with modification was the origin of *all* living things after the first organisms. He wrote in *The Origin of Species:* "I view all beings not as special creations, but as the lineal descendants of some few beings" that lived in the distant past. The reason living things are now so different from each other, Darwin believed, is that they have been modified by natural selection, or survival of the fittest: "I am convinced that Natural Selection has been the most important, but not the exclusive, means of modification."

When proponents of Darwin's theory are responding to critics, they sometimes claim that "evolution" means simply change over time. But this is clearly an evasion. No rational person denies the reality of change, and we did not need Charles Darwin to convince us of it. If "evolution" meant only this, it would be utterly uncontroversial. Nobody believes that biological evolution is simply change over time.

Only slightly less evasive is the statement that descent with modification occurs. Of course it does, because all organisms within a single species are related through descent with modification. We see this in our own families, and plant and animal breeders see it in their work. But this still misses the point.

No one doubts that descent with modification occurs in the course of ordinary biological reproduction. The question is whether descent with modification accounts for the origin of *new* species—in fact, of *every* species. Like change over time, descent with modification within a species is utterly uncontroversial. But Darwinian evolution claims much more. In particular, it claims that descent with modification explains the origin and diversification of *all* living things.

The only way anyone can determine whether this claim is true is by comparing it with observations or experiments. Like all other scientific theories, Darwinian evolution must be continually compared with the evidence. If it does not fit the evidence, it must be reevaluated or abandoned—otherwise it is not science, but myth.

Evidence for evolution

When asked to list the evidence for Darwinian evolution, most people—including most biologists—give the same set of examples, because all of them learned biology from the same few textbooks. The most common examples are:

- a laboratory flask containing a simulation of the Earth's primitive atmosphere, in which electric sparks produce the chemical building-blocks of living cells;
- the evolutionary tree of life, reconstructed from a large and growing body of fossil and molecular evidence;
- similar bone structures in a bat's wing, a porpoise's flipper, a horse's leg, and a human hand that indicate their evolutionary origin in a common ancestor;
- pictures of similarities in early embryos showing that amphibians, reptiles, birds and human beings are all descended from a fish-like animal;
- *Archaeopteryx*, a fossil bird with teeth in its jaws and claws on its wings, the missing link between ancient reptiles and modern birds;
- peppered moths on tree trunks, showing how camouflage and predatory birds produced the most famous example of evolution by natural selection;
- Darwin's finches on the Galápagos Islands, thirteen separate species that diverged from one when natural selection produced differences in their beaks, and that inspired Darwin to formulate his theory of evolution;
- fruit flies with an extra pair of wings, showing that genetic mutations can provide the raw materials for evolution;
- a branching-tree pattern of horse fossils that refutes the old-fashioned idea that evolution was directed; and

- drawings of ape-like creatures evolving into humans, showing that we are just animals and that our existence is merely a by-product of purposeless natural causes.

These examples are so frequently used as evidence for Darwin's theory that most of them have been called "icons" of evolution. Yet all of them, in one way or another, misrepresent the truth.

Science or myth?

Some of these icons of evolution present assumptions or hypotheses as though they were observed facts; in Stephen Jay Gould's words, they are "incarnations of concepts masquerading as neutral descriptions of nature." Others conceal raging controversies among biologists that have far-reaching implications for evolutionary theory. Worst of all, some are directly contrary to well-established scientific evidence.

Most biologists are unaware of these problems. Indeed, most biologists work in fields far removed from evolutionary biology. Most of what they know about evolution, they learned from biology textbooks and the same magazine articles and television documentaries that are seen by the general public. But the textbooks and popular presentations rely primarily on the icons of evolution, so as far as many biologists are concerned the icons *are* the evidence for evolution.

Some biologists are aware of difficulties with a particular icon because it distorts the evidence in their own field. When they read the scientific literature in their specialty, they can see that the icon is misleading or downright false. But they may feel that this is just an isolated problem, especially when they are assured that Darwin's theory is supported by overwhelming evidence from other fields. If they believe in the fundamental correctness of Darwinian evolution, they may set aside their misgivings about the particular icon they know something about.

On the other hand, if they voice their misgivings they may find it difficult to gain a hearing among their colleagues, because (as we shall see) criticizing Darwinian evolution is extremely unpopular among English-speaking biologists. This may be why the problems with the icons of evolution are not more widely known. And this is why many biologists will be just as surprised as the general public to learn how serious and widespread those problems are.

The following chapters compare the icons of evolution with published scientific evidence, and reveal that much of what we teach about evolution is wrong. This fact raises troubling questions about the status of Darwinian evolution. If the icons of evolution are supposed to be our best evidence for Darwin's theory, and all of them are false or misleading, what does that tell us about the theory? Is it science, or myth?

CHAPTER 2

The Miller–Urey Experiment

Accompanied by music from Stravinsky's *Rite of Spring*, the primordial Earth seethes with volcanic activity. Red-hot lava flows over the land and tumbles into the sea, generating clouds of steam while lightning flashes in the sky above. Slowly, the camera pans down until it reaches the calm depths of the ocean, where mysterious specks glow in the dark. Suddenly, a single-celled animal darts across the screen. Life is born.

The scene is from Walt Disney's 1940 film classic, *Fantasia,* and the narrator calls it "a coldly accurate reproduction of what science thinks went on during the first few billion years of this planet's existence." The scenario was the brain-child of Russian scientist A. I. Oparin and British scientist J. B. S. Haldane, who in the 1920s had suggested that lightning in the primitive atmosphere could have produced the chemical building blocks of life. Although Darwin did not pretend to understand the origin of life, he speculated that it might have started in "some warm little pond." Similarly, Oparin and Haldane hypothesized that chemicals produced in the atmosphere dissolved in the primordial seas to form a "hot dilute soup," from which the first living cells emerged.

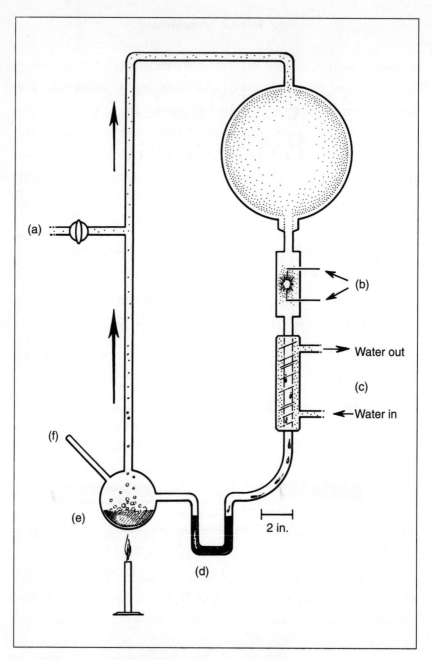

FIGURE 2-1 The 1953 Miller-Urey Experiment.

FIGURE 2-1   The 1953 Miller-Urey experiment.

(a) Vacuum line; (b) high-voltage spark electrodes; (c) condenser with circulating cold water; (d) trap to prevent backflow; (e) flask for boiling water and collecting reaction products; (f) sealed tube, broken later to remove reaction products for analysis. In later experiments, the electrodes were moved up into the large flask at the upper right, and a stopcock for withdrawing reaction products was added to the trap at the bottom. Most textbook drawings show these later modifications.

The Oparin-Haldane hypothesis captured the imagination of many scientists, and thus became "what science thinks" about the first steps in the origin of life. But it remained an untested hypothesis until the early 1950s, when an American graduate student, Stanley Miller, and his Ph.D. advisor, Harold Urey, produced some of the chemical building blocks of life by sending an electric spark through a mixture of gases they thought simulated the Earth's primitive atmosphere. (Figure 2-1) The 1953 Miller-Urey experiment generated enormous excitement in the scientific community, and soon found its way into almost every high school and college biology textbook as evidence that scientists had demonstrated the first step in the origin of life.

The Miller-Urey experiment is still featured prominently in textbooks, magazines, and television documentaries as an icon of evolution. Yet for more than a decade most geochemists have been convinced that the experiment failed to simulate conditions on the early Earth, and thus has little or nothing to do with the origin of life. Here's why.

The Oparin-Haldane scenario

The first step in the Oparin-Haldane scenario—the production of life's chemical building blocks by lightning—depends crucially on the composition of the atmosphere. The Earth's present atmosphere is about 21 percent oxygen gas. We tend to think of an oxygen-rich atmosphere as essential to life, because we would die without it. Yet, paradoxically, life's building blocks could not have formed in such an atmosphere.

We need oxygen because our cells produce energy through aerobic respiration (though some bacteria are "anaerobic," and thrive in the absence of oxygen). In effect, aerobic organisms use oxygen to get energy from organic molecules in much the same way that automobile engines use oxygen to get energy from gasoline. But our bodies must also synthesize organic molecules, otherwise we could not grow, heal, or reproduce. Respiration, which breaks down organic molecules, is the opposite of synthesis, which builds them up. Chemists call the process of respiration "oxidizing," while they call the process of synthesis "reducing."

Not surprisingly, the same oxygen that is essential to aerobic respiration is often fatal to organic synthesis. An electric spark in a closed container of swamp gas (methane) might produce some interesting organic molecules, but if even a little oxygen is present the spark will cause an explosion. Just as a closed container excludes oxygen and prevents swamp gas from exploding, so compartments in living cells exclude oxygen from the processes of organic synthesis. Free oxygen in the wrong places can be harmful to health, which is why some nutritionists tell people to consume more "anti-oxidant" vitamins.

Since free oxygen can destroy many organic molecules, chemists often must remove oxygen and use closed containers when they synthesize and store organic chemicals in the laboratory. But before the origin of life, when there were neither chemists nor laboratories, the chemical building blocks of life could have formed only in a natural environment lacking oxygen. According to Oparin and Haldane, that environment was the Earth's primitive atmosphere.

The Earth's present atmosphere is strongly oxidizing. Oparin and Haldane postulated its exact opposite: a strongly reducing atmosphere rich in hydrogen. Specifically, they postulated a mixture of methane (hydrogen combined with carbon), ammonia (hydrogen combined with nitrogen), water vapor (hydrogen combined with oxygen) and hydrogen gas. Oparin and Haldane predicted that lightning in such an atmosphere could spontaneously produce the organic molecules needed by living cells.

The Miller-Urey experiment

At the time, it seemed reasonable to postulate a strongly reducing primitive atmosphere. Scientists believed that the Earth originally formed from a condensing cloud of interstellar dust and gas, so it was reasonable to suppose that the original atmosphere resembled interstellar gases, which consist predominantly of hydrogen. In 1952, Nobel Prize-winning chemist Harold Urey concluded that the early atmosphere consisted primarily of hydrogen, methane, ammonia and water vapor—just as Oparin and Haldane had postulated.

Urey's graduate student at The University of Chicago, Stanley Miller, set out to test the Oparin-Haldane hypothesis experimentally. Miller assembled a closed glass apparatus in Urey's laboratory, pumped out the air, and replaced it with methane, ammonia, hydrogen and water. (If he hadn't removed the air, his next step might have been his last.) He then heated the water and circulated the gases past a high-voltage electric spark to simulate lightning. (Figure 2-1)

"By the end of the week," Miller reported, the water "was deep red and turbid." He removed some of it for chemical analysis and identified several organic compounds. These included glycine and alanine, the two simplest amino acids found in proteins. Most of the reaction products, however, were simple organic compounds that do not occur in living organisms.

Miller published his initial results in 1953. By repeating the experiment, he and others were able to obtain small yields of most biologically significant amino acids, as well as some additional organic compounds found in living cells. The Miller-Urey experiment thus seemed to confirm the Oparin-Haldane hypothesis about the first step in the origin of life. By the 1960s, however, geochemists were beginning to doubt that conditions on the early Earth were the ones Oparin and Haldane had postulated.

Did the primitive atmosphere really lack oxygen?

Urey assumed that the Earth's original atmosphere had the same composition as interstellar gas clouds. In 1952, however (the same year Urey published this view), University of Chicago geochemist Harrison Brown noted that the abundance of the rare gases neon, argon, krypton, and xenon in the Earth's atmosphere was at least a million times lower than the cosmic average, and concluded that the Earth must have lost its original atmosphere (if it ever had one) very soon after its formation.

In the 1960s Princeton University geochemist Heinrich Holland and Carnegie Institution geophysicist Philip Abelson agreed with Brown. Holland and Abelson independently concluded that the Earth's primitive atmosphere was *not* derived from interstellar gas clouds, but from gases released by the Earth's own volcanoes. They saw no reason to believe that ancient volcanoes were different from modern ones, which release primarily water vapor, carbon dioxide, nitrogen, and trace amounts of hydrogen. Since hydrogen is so light, Earth's gravity would have been unable to hold it, and (like the rare gases) it would quickly have escaped into space.

But if the principal ingredient of the primitive atmosphere was water vapor, the atmosphere must also have contained some oxygen. Atmospheric scientists know that ultraviolet rays from sunlight cause dissociation of water vapor in the upper atmosphere. This process, called "photodissociation," splits water molecules into hydrogen and oxygen. The hydrogen escapes into space, leaving the oxygen behind in the atmosphere. (Figure 2-2)

Scientists believe that most of the oxygen in the present atmosphere was produced later by photosynthesis, the process by which green plants convert carbon dioxide and water into organic matter and oxygen.

Nevertheless, photodissociation would have generated small amounts of oxygen even before the advent of photosynthesis. The question is, how much?

In 1965 Texas scientists L. V. Berkner and L. C. Marshall argued that the oxygen produced by photodissociation could not have exceeded about one thousandth of its present atmospheric level, and was probably much lower. California Institute of Technology geophysicist R. T. Brinkmann disagreed, claiming that "appreciable oxygen concentrations might have evolved in the Earth's atmosphere"—as much as one quarter of the present level—before the advent of photosynthesis. As the controversy over theoretical implications widened, various scientists took one side or the other: Australian geologist J. H. Carver concurred with Brinkmann, while Pennsylvania State University geologist James Kasting agreed with Berkner and Marshall. The issue was never resolved.

Evidence from ancient rocks has been inconclusive. Some ancient sedimentary rocks contain uraninite, an oxygen-poor uranium compound that suggests to some geologists that those sediments had been laid down in an oxygen-poor atmosphere. But other geologists point out that uraninite also occurs in later rocks that were deposited in our modern oxygen-rich atmosphere. Sediments rich in the highly oxidized red form of iron have also been used to infer primitive oxygen levels. Geologist James C. G. Walker argued that the appearance of these "red-beds" about two billion years ago "marks the beginning of the aerobic atmosphere." But red-beds also occur in rocks older than two billion years, and Canadian geologists Erich Dimroth and Michael Kimberly wrote in 1979 that they saw "no evidence" in the sedimentary distribution of iron "that an oxygen-free atmosphere has existed at any time during the span of geological history recorded in well preserved sedimentary rocks."

Biochemical evidence has been used to infer primitive oxygen levels, as well. In 1975 British biologists J. Lumsden and D. O. Hall reported that an enzyme (superoxide dismutase) used by living cells to protect themselves from the damaging effects of oxygen is present even in organisms whose ancestors are thought to have existed before the advent of photosynthesis. Lumsden and Hall concluded that the enzyme must have evolved to provide protection against primitive oxygen produced by photo-dissociation.

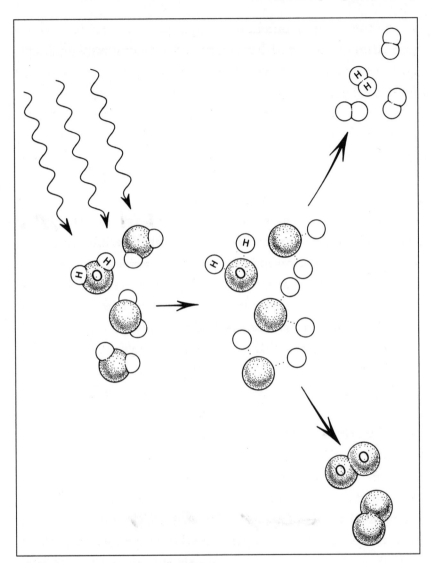

FIGURE 2-2   Photodissociation.

Water molecules (oxygen plus hydrogen) are split by ultraviolet rays from sunlight in the upper atmosphere. The hydrogen (H) is too light to be held by Earth's gravity and escapes into outer space, while the heavier oxygen (O) remains in the atmosphere.

So theoretical models implied some primitive oxygen, but no one knew how much. Evidence from the rocks was inconclusive, and the biochemical evidence seemed to point to significant levels of oxygen produced by photodissociation. The controversy raged from the 1960s until the early 1980s, when it faded from view.

Declaring the controversy over

In 1977 origin-of-life researchers Sidney Fox and Klaus Dose reported that a major reason why the Earth's primitive atmosphere "is widely believed not to have contained in its early stage significant amounts of oxygen" is that "laboratory experiments show that chemical evolution, as accounted for by present models, would be largely inhibited by oxygen." James C. G. Walker likewise wrote that "the strongest evidence" for the composition of the primitive atmosphere "is provided by conditions for the origin of life. A reducing atmosphere is required."

Participants at a 1982 conference on the origin of life (one of whom was Stanley Miller) agreed that there could not have been free oxygen in the early atmosphere "because reducing conditions are required for the synthesis of the organic compounds needed for the development of life." That same year, British geologists Harry Clemmey and Nick Badham wrote that the evidence showed "from the time of the earliest dated rocks at 3.7 billion years ago, Earth had an oxygenic atmosphere." Clemmey and Badham declared it a mere "dogma" to claim that the Earth's early atmosphere lacked oxygen.

But geological and biochemical evidence no longer mattered, because certain influential scientists decided that the Miller-Urey experiment had demonstrated the first step in the origin of life, and they simply declared that the primitive atmosphere must have lacked oxygen. Clemmey and Badham were right. Dogma had taken the place of empirical science.

From a scientific perspective, this dogma puts the cart before the horse. The Miller-Urey experiment succeeded in synthesizing organic molecules, but the question was not whether organic molecules could be synthesized in the laboratory. Of course they could, and they had been for years. They can be synthesized in the laboratory even though the present atmosphere is strongly oxidizing, because chemists create local environments from which oxygen is excluded or maintained at extremely low levels. The success of the Miller-Urey experiment doesn't prove that the entire primitive atmosphere lacked oxygen any more than the success of modern organic chemistry proves that the modern atmosphere lacks oxygen.

Clearly, some of the geological and biochemical evidence points to oxygen in the primitive atmosphere; otherwise, the issue would not have been so hotly debated among geologists from the 1960s through the early 1980s. In fact, evidence for primitive oxygen continues to mount: Smithsonian Institution paleobiologist Kenneth Towe (now emeritus) reviewed the evidence in 1996, and concluded that "the early Earth very likely had an atmosphere that contained free oxygen."

The evidence Towe cited is usually ignored by people currently involved in origin-of-life research, and has been for years. Ironically, however, not even this arbitrary dismissal of evidence saved the Miller-Urey experiment. Although geochemists were sharply divided on the oxygen issue, they soon reached a near-consensus that the primitive atmosphere was nothing like the one Miller used.

The Miller-Urey experiment fails anyway

Holland and Abelson concluded in the 1960s that the Earth's primitive atmosphere was derived from volcanic outgassing, and consisted primarily of water vapor, carbon dioxide, nitrogen, and trace amounts of hydrogen. With most of the hydrogen being lost to space, there would have been nothing to reduce the carbon dioxide and nitrogen, so methane and ammonia could not have been major constituents of the early atmosphere. (Figure 2-3)

Abelson also noted that ammonia absorbs ultraviolet radiation from sunlight, and would have been rapidly destroyed by it. Furthermore, if large amounts of methane had been present in the primitive atmosphere, the earliest rocks would have contained a high proportion of organic molecules, and this is not the case. Abelson concluded: "What is the evidence for a primitive methane-ammonia atmosphere on Earth? The answer is that there is *no* evidence for it, but much against it." (emphasis in original) In other words, the Oparin-Haldane scenario was wrong, and the early atmosphere was nothing like the strongly reducing mixture used in Miller's experiment.

Other scientists agreed. In 1975 Belgian biochemist Marcel Florkin announced that "the concept of a reducing primitive atmosphere has been abandoned," and the Miller-Urey experiment is "not now considered geologically adequate." Sidney Fox and Klaus Dose—though they argued that the primitive atmosphere lacked oxygen—conceded in 1977 that a reducing atmosphere did "not seem to be geologically realistic because evidence indicates that... most of the free hydrogen probably had disappeared into outer space and what was left of methane and ammonia was oxidized."

According to Fox and Dose, not only did the Miller-Urey experiment start with the wrong gas mixture, but also it did "not satisfactorily represent early geological reality because no provisions [were] made to remove hydrogen from the system." During a Miller-Urey experiment hydrogen gas accumulates, becoming up to 76 percent of the mixture, but on the early Earth it would have escaped into space. Fox and Dose concluded: "The inference that Miller's synthesis does not have a geological relevance has become increasingly widespread."

Since 1977 this view has become a near-consensus among geochemists. As Jon Cohen wrote in *Science* in 1995, many origin-of-life researchers now dismiss the 1953 experiment because "the early atmosphere looked nothing like the Miller-Urey simulation."

So what? Maybe a water vapor–carbon dioxide–nitrogen atmosphere would still support a Miller–Urey-type synthesis (as long as oxygen is excluded). But Fox and Dose reported in 1977 that no amino acids are produced by sparking such a mixture, and Heinrich Holland noted in 1984 that the "yields and the variety of organic compounds produced in these experiments decrease considerably" as methane and ammonia are removed from the starting mixtures. According to Holland, mixtures of carbon dioxide, nitrogen, and water yielded no amino acids at all.

In 1983 Miller reported that he and a colleague were able to produce a small amount of the simplest amino acid, glycine, by sparking an atmosphere containing carbon monoxide and carbon dioxide instead of methane, as long as free hydrogen was present. But he conceded that glycine was about the best they could do in the absence of methane. As John Horgan wrote in *Scientific American* in 1991, an atmosphere of carbon dioxide, nitrogen,

OXIDIZING (present Earth)	NEUTRAL (volcanic outgassing)	REDUCING (Oparin-Haldane)
nitrogen	water vapor (hydrogen + oxygen)	methane (carbon + hydrogen)
oxygen	carbon dioxide (carbon + oxygen)	ammonia (nitrogen + hydrogen)
carbon dioxide (carbon + oxygen)	nitrogen	hydrogen
water vapor (hydrogen + oxygen)	hydrogen (trace; lost to space)	water vapor (oxygen + hydrogen)

FIGURE 2-3 A comparison of oxidizing, neutral, and reducing atmospheres

Constituents are listed from top to bottom in order of their prevalance.

and water vapor "would not have been conducive to the synthesis of amino acids."

The conclusion is clear: if the Miller–Urey experiment is repeated using a realistic simulation of the Earth's primitive atmosphere, it doesn't work. Therefore, origin–of–life researchers have had to look elsewhere.

An RNA world?

Since the Miller–Urey experiment fails to explain how proteins could have formed on the early Earth, origin-of-life researchers have considered the possibility that proteins were not the first molecular building-blocks of life. DNA is not a good candidate, because it needs a whole suite of complex proteins to make copies of itself. Therefore DNA could not have originated before proteins, and could not have been the first step in the origin of life.

Another candidate is RNA, a close chemical relative of DNA that is used by all living cells in the process of making proteins. In the 1980s molecular biologists Thomas Cech and Sidney Altman showed that RNA can sometimes behave like an enzyme—that is, like a protein. Another molecular biologist, Walter Gilbert, suggested that RNA might be able to synthesize itself in the absence of proteins, and thus might have originated on the early Earth before either proteins or DNA. This "RNA world" might then have been the molecular cradle from which living cells emerged.

But no one has demonstrated how RNA could have formed before living cells were around to make it. According to Scripps Research Institute biochemist Gerald Joyce, RNA is not a plausible candidate for the first building block of life "because it is unlikely to have been produced in significant quantities on the primitive Earth." Even if RNA could have been produced, it would not have survived long under the conditions thought to have existed on the early Earth.

Joyce concludes: "The most reasonable interpretation is that life did not start with RNA." Although he still thinks that an RNA world preceded the DNA world, he believes that some kind of living cells must have preceded RNA. "You have to build straw man upon straw man," Joyce said in 1998, "to get to the point where RNA is a viable first biomolecule."

In other words, the RNA world—like the protein-first scenario in the Miller-Urey experiment—is a dead end. Origin-of-life researchers have been unable to show how the molecular building blocks of life formed on the early Earth. But even if they had discovered the origin of the building blocks, the origin of life would remain a mystery. A biochemist can mix all the chemical building blocks of life in a test tube and still not produce a living cell.

The origin of life problem is so difficult that German researcher Klaus Dose wrote in 1988 that current theory is "a scheme of ignorance. Without fundamentally new insights in evolutionary processes… this ignorance is likely to persist." And persist it has. In 1998, comparing the scientific search for the origin of life to a detective story, Salk Institute scientist Leslie Orgel acknowledged that "we are very far from knowing whodunit." And *New York Times* science writer Nicholas Wade reported in June 2000: "Everything about the origin of life on Earth is a mystery, and it seems the more that is known, the more acute the puzzles get."

So we remain profoundly ignorant of how life originated. Yet the Miller-Urey experiment continues to be used as an icon of evolution, because nothing better has turned up. Instead of being told the truth, we are given the misleading impression that scientists have empirically demonstrated the first step in the origin of life.

The Miller-Urey experiment as an icon of evolution

The March 1998 issue of *National Geographic* carries a photo of Miller standing next to his experimental apparatus. The caption reads: "Approximating conditions on the early Earth in a 1952 experiment, Stanley Miller—now at the University of California at San Diego—produced amino acids. 'Once you get the equipment together it's very simple,' he says."

Several pages later, the *National Geographic* article explains: "Many scientists now suspect that the early atmosphere was different from what Miller first supposed." But a picture is worth a thousand words—especially when its caption is misleading and the truth is buried deep in the article. Even a careful reader is left with the impression that the Miller-Urey experiment showed how easy it was for life to originate on the early Earth.

Many biology textbooks use the same misleading approach. The 2000 edition of Kenneth Miller and Joseph Levine's *Biology,* a popular high school textbook, includes a drawing of the Miller-Urey apparatus with the caption: "By re-creating the early atmosphere (ammonia, water, hydrogen and methane) and passing an electric spark (lightning) through the mixture, Miller and Urey proved that organic matter such as amino acids could have formed spontaneously." Like the *National Geographic* article, the Miller-Levine textbook buries a disclaimer in the text: "Miller's original guesses about the Earth's early atmosphere were probably incorrect," but even this is softened by adding that experiments using other mixtures "also have produced organic compounds." In any case, the textbook is quite adamant that the ancient atmosphere "did not contain oxygen gas."

The 1998 college textbook, *Life: The Science of Biology* by William Purves, Gordon Orians, Craig Heller, and David Sadava, tells students that Stanley Miller produced "the building blocks of life" using "a reducing atmosphere such as existed on early Earth," and that "no free oxygen was present in this early atmosphere." This textbook gives students no hint that most scientists now think the Miller-Urey experiment failed to simulate actual conditions on the early Earth.

Even advanced college textbooks misrepresent the truth. The 1998 edition of Douglas Futuyma's *Evolutionary Biology* includes a drawing of "the apparatus Miller used to synthesize organic molecules under simulated early Earth conditions." The only thing Futuyma's book has to say about the controversy over primitive oxygen is that "at the time of the earliest life, the atmosphere virtually lacked oxygen." And the latest edition of *Molecular Biology of the Cell*, a graduate level textbook by National Academy of Sciences President Bruce Alberts and his colleagues, features the Miller-Urey apparatus and calls it "a typical experiment simulating conditions on the primitive Earth." The accompanying text asserts that organic molecules "are likely to have been produced under such conditions. The best evidence for this comes from laboratory experiments."

A 1999 booklet published by the National Academy of Sciences perpetuates the misrepresentation: "Experiments conducted under conditions intended to resemble those present on primitive Earth have resulted in the production of some of the chemical components of proteins." This booklet includes a preface by Bruce Alberts, who (as we saw in the Introduction) assures us that "science and lies cannot coexist."

This is even more troubling than the misuse of the Miller-Urey experiment by *National Geographic* and biology textbooks. The National Academy of Sciences is the nation's premier science organization, commissioned by Congress in 1863 to advise the government on scientific matters. Its members include many of the best scientists in America. Do they really approve of misleading the public about the evidence for evolution? Or is this being done without the members' knowledge? What are the American people supposed to think?

As we shall see in the following chapters, booklets published recently by the National Academy contain other false and misleading statements about evolution, too. Clearly, we are not dealing here with an isolated textbook error. The implications for American science are potentially far-reaching.

In 1986 chemist Robert Shapiro wrote a book criticizing several aspects of research on the origin of life. He was especially critical of the argument that the Miller-Urey experiment proved that the Earth's primitive atmosphere was strongly reducing. "We have reached a situation," he wrote, "where a theory has been accepted as fact by some, and possible contrary evidence is shunted aside." He concluded that this is "mythology rather than science."

Are we teaching our biology students mythology rather than science?

CHAPTER 3

Darwin's Tree of Life

No one knows how the first living cells originated, but most biologists think the event was so improbable that it happened only once—or, at most, a few times. If so, then it is reasonable to suppose that those few original cells gave rise to the millions of different species alive today. This was Charles Darwin's view in *The Origin of Species*: "I view all beings not as special creations, but as the lineal descendants of some few beings which lived long before the first bed of the Cambrian system was deposited." (When Darwin wrote *The Origin of Species* in 1859, the Cambrian was the oldest geological period in which fossils had been found.) Indeed, Darwin thought that "all the organic beings which have ever lived on this earth may be descended from some one primordial form."

The Origin of Species included only one illustration, showing the branching pattern that would result from this process of descent with modification. (Figure 3-1) Darwin thus pictured the history of life as a tree, with the universal common ancestor as its root and modern species as its "green and budding twigs." He called this the "great Tree of Life."

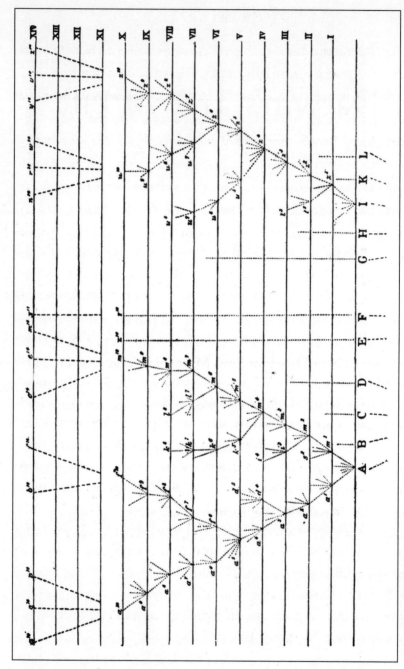

FIGURE 3-1 Darwin's tree of life.

FIGURE 3-1 Darwin's tree of life.

The only illustration in Darwin's *Origin of Species*, showing the branching pattern of divergence predicted by his theory. The vertical dimension represents time, with oldest at the bottom and most recent at the top, while the horizontal dimension represents degrees of differences among organisms. As the illustration shows, Darwin conceived of some lineages as persisting largely unchanged. The dotted lines at the bottom reflect Darwin's conviction that the eleven lineages shown here descended from still fewer lineages. Indeed, he believed that "one primordial form" may have been the common ancestor of all living things. Most of Darwin's modern followers believe that the origin of life was sufficiently improbable that the tree of life is rooted in a single universal common ancestor.

Of all the icons of evolution, the tree of life is the most pervasive, because descent from a common ancestor is the foundation of Darwin's theory. Neo-Darwinist Ernst Mayr boldly proclaimed in 1991 that "there is probably no biologist left today who would question that all organisms now found on the earth have descended from a single origin of life." Yet Darwin knew—and scientists have recently confirmed—that the early fossil record turns the evolutionary tree of life upside down. Ten years ago it was hoped that molecular evidence might save the tree, but recent discoveries have dashed that hope. Although you would not learn it from reading biology textbooks, Darwin's tree of life has been uprooted.

Darwin's tree of life

If all living things are descended from a common ancestor, why are they so different? Domestic breeders modify existing stocks by selecting only certain variants for breeding. Darwin argued that an analogous process operates in the wild. If part of a natural population were exposed to one set of conditions, and other parts were exposed to other conditions, "natural selection" could modify the various sub-populations in different ways. Given enough time, one species could produce several varieties; and Darwin believed that if those varieties continued to diverge, they would eventually become separate species.

In the system of biological classification invented by Carolus Linnaeus a century before Darwin (and still used by most biologists), organisms are grouped on the basis of similarities and differences into a hierarchical series. The *species* is the lowest level of the hierarchy; *genus* (plural "genera") is the next, then *family*, *order, class, phylum* (plural "phyla"; called a "division" in plants and fungi), and the highest level, *kingdom*. For example, the species name for human beings is *sapiens*, and the genus name is *Homo*; both are included in the scientific name, which is *Homo sapiens*. Humans are grouped together with apes in the Hominid family; hominids and monkeys are grouped together in the Primate order, then grouped with other warm-blooded, milk-producing animals in the Mammal class. Mammals, in turn, are placed in the Chordate phylum (the "chord" is an embryonic structure that in most members of this phylum becomes a backbone; such animals are called "vertebrates"). At the highest level of the hierarchy, the Animal kingdom includes several dozen phyla.

For comparison, the common fruit fly is called *Drosophila melanogaster* (genus and species). It is a member of the Drosophilid family, which is grouped with other two-winged insects in the Diptera order, and these are grouped with other six-legged animals in the Insect class. Insects are grouped with other organisms possessing external skeletons and jointed appendages (lobsters, for example) in the Arthropod phylum, which (like the Chordate phylum) is in the Animal kingdom. (Other kingdoms include plants, fungi, and bacteria.) (Figure 3-2)

According to Darwin's theory, humans and fruit flies shared a common ancestor (which probably looked nothing like humans or fruit flies) sometime in the distant past. Darwin believed that if we could have been there to observe the process, we would have seen the ancestral species split into several species only slightly different from each other. These species would then have evolved in different directions under the influence of natural selection. More and more distinct species would have appeared; and eventually, at least one of them would have become so different from the others that it could be considered a different genus. As generations passed, differences would have continued to accumulate, eventually giving rise to separate families.

	Humans	Fruit Flies
KINGDOM	Animals	Animals
PHYLUM	Chordates	Arthropods
CLASS	Mammals	Insects
ORDER	Primates	Diptera
FAMILY	Hominids	Drosophilids
GENUS	*Homo*	*Drosophila*
SPECIES	*sapiens*	*melanogaster*

FIGURE 3-2 **Biological classification.**

Devised by Carolus Linnaeus a century before Darwin, the Linnaean system classifies organisms into increasingly more inclusive groups. Only the major categories are shown here; there are also intermediate categories such as "sub-phylum Vertebrates" (animals with backbones, which comprise most of the Chordates).

This was the process Darwin illustrated in *The Origin of Species*. (Figure 3-1) The vertical dimension in Darwin's drawing represents time, with oldest at the bottom and newest at the top, while the horizontal dimension represents differences among organisms. Darwin believed that minor variations within the original ancestral species were gradually amplified over the course of many generations into larger differences that separated species from one another. As he put it, "the small differences distinguishing varieties of the same species, steadily tend to increase, till they equal the greater differences between species."

Taking each horizontal line in his illustration to indicate a thousand generations, Darwin estimated that "six new species, marked by the letters n^{14} to z^{14}" at the top, might have been produced after fourteen thousand generations. In fact, since "the original species (I) differed largely from (A), standing nearly at the extreme end of the original genus" at the bottom, it seemed probable that "the six new species descended from (I), and the eight descendants from (A), will have to be ranked as very distinct genera, or even as distinct sub-families."

Still greater differences could be explained on a larger time scale. For example, if one were to take "each horizontal line [to] represent a million or more generations," Darwin saw "no reason to limit the process of modification, as now explained, to the formation of genera alone," but considered it equally capable of producing "new families, or orders,... [or] classes." Thus the large differences separating orders and classes would emerge only after a very long history of small differences: "As natural selection acts solely by accumulating slight, successive, favorable variations, it can produce no great or sudden modifications; it can act only by short and slow steps." These "short and slow steps" give Darwin's illustration its characteristic branching-tree pattern.

Therefore, if the bottom line in Darwin's diagram represents varieties, the top line might be different species or genera. If we take those genera, put them at the bottom, and start the process over, we might get families or orders; then if we put those orders at the bottom and repeat the process, we might get classes or even phyla. But in Darwin's theory, there is no way phylum-level differences could have appeared right at the start. Yet that is what the fossil record shows.

Darwin and the fossil record

When Darwin wrote *The Origin of Species*, the oldest known fossils were from a geological period known as the Cambrian, named after rocks in Cambria, Wales. (Figure 3-3) But the Cambrian fossil pattern didn't fit Darwin's theory. Instead of starting with one or a few species that diverged gradually over millions of years into families, then orders, then classes, then phyla, the Cambrian starts with the abrupt appearance of many fully-formed phyla and classes of animals. In other words, the highest levels of the biological hierarchy appeared right at the start.

Darwin was aware of this, and considered it a major difficulty for his theory. He wrote in *The Origin of Species* that "if the theory be true, it is indisputable that before the lowest Cambrian stratum was deposited long periods elapsed... [in which] the world swarmed with living creatures." Yet he acknowledged that "several of the main divisions of the animal kingdom suddenly appear in the lowest known fossiliferous rocks." Darwin called this a "serious" problem which "at present must remain inexplicable; and may be truly urged as a valid argument against the views here entertained."

Darwin was convinced, however, that the difficulty was only apparent. The fossil record is "a history of the world imperfectly kept," he argued, "and written in a changing dialect; of this history

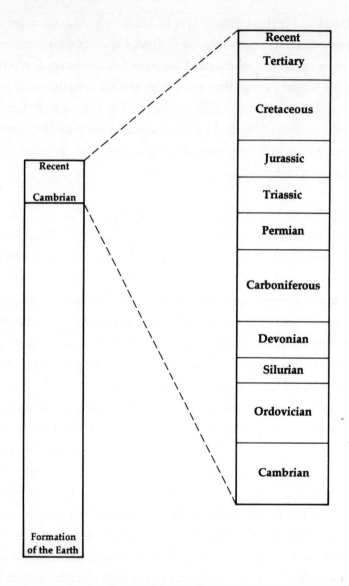

FIGURE 3-3 The Geological record.

The column on the left represents the entire history of the earth since its formation, currently dated at about four and a half billion years ago. The column on the right represents slightly more than the last ten percent of this.

we possess the last volume alone, relating only to two or three countries." He believed that rocks older than the Cambrian period had been so altered by heat and pressure as to destroy all vestiges of fossils; because of this, the major groups of animals only "falsely appear to have been abruptly introduced" in the Cambrian. Darwin also pointed out that "only a small portion of the surface of the earth has been geologically explored," as if to suggest that further fossil-hunting might provide at least some of the missing evidence.

Since that time, further exploration has turned up many fossil beds older than the Cambrian, so our present understanding of Precambrian history is far better than Darwin's. Paleontologists have also found Cambrian rocks in Canada, Greenland, and China where well-preserved fossils are particularly plentiful. But this vastly improved knowledge of Cambrian and Precambrian fossils has aggravated Darwin's problem rather than alleviated it. Many paleontologists are now convinced that the major groups of animals really *did* appear abruptly in the early Cambrian. The fossil evidence is so strong, and the event so dramatic, that it has become known as "the Cambrian explosion," or "biology's big bang."

The Cambrian explosion

In Africa and Australia, geologists have reported unmetamorphosed sediments more than three billion years old that contain fossilized single-celled organisms. Sediments only slightly younger have been found that contain fossil "stromatolites," layered mats of photosynthetic bacteria and sediment that form in shallow seas. But Precambrian fossils consisted only of single-celled organisms until just before the Cambrian.

Multicellular organisms slightly older than the Cambrian were first discovered in the Ediacara Hills in South Australia, but are now known from many other locations around the world. Some paleontologists argue that the Ediacaran fossils were ancestors of the animals that appeared later in the Cambrian, while others claim they are so utterly different from all other life-forms that they should be placed in their own kingdom. British paleontologist Simon Conway Morris believes that at least some of the Ediacaran fossils were animals, but maintains that most of the many species appearing in the Cambrian did not have ancestors in Ediacara. "Apart from the few Ediacaran survivors," wrote Conway Morris in 1998, "there seems to be a sharp demarcation between the strange world of Ediacaran life and the relatively familiar Cambrian fossils."

There are two other indications of multicellular animals just before the Cambrian: a "small shelly fauna," consisting of tiny fossils that are unlike any modern group, and trace fossils (burrows and tracks), apparently left by multicellular worms. But except for the latter, and possibly a few survivors from Ediacara, there is no fossil evidence connecting Cambrian animals to organisms that preceded them. The now well-documented Precambrian fossil record does not provide anything like the long history of gradual divergence required by Darwin's theory.

Although the abrupt appearance of animal fossils in the Cambrian was known to Darwin, the full extent of the phenomenon wasn't appreciated until the 1980s, when fossils from the previously-discovered Burgess Shale in Canada were re-analyzed by paleontologists Harry Whittington, Derek Briggs and Simon Conway Morris. The 1980s also marked the discovery of two other fossil locations similar to the Burgess

Shale: the Sirius Passet in northern Greenland, and the Chengjiang in southern China. All of these locations document the bewildering variety of animals that appeared in the Cambrian. The Chengjiang fossils, however, appear to be the earliest and best-preserved, and they include what may be the first vertebrates.

Various dates have been proposed for the Cambrian period and the time of the Cambrian explosion, with recent estimates ranging between 600 and 500 million years ago. In 1993 geologist Samuel Bowring and his colleagues summarized the available evidence from the rock strata and radioactive dating methods, and concluded that the Cambrian period began about 544 million years ago. The major increase in animal fossils that marks the Cambrian explosion began about 530 million years ago, and lasted a maximum of 5 to 10 million years. (Although 10 million years is a long time in human terms, it is short in geological terms, amounting to less than 2 percent of the time elapsed since the beginning of the Cambrian.) The Cambrian explosion gave rise to most of the animal phyla alive today, as well as some that are now extinct. (Figure 3-4)

According to paleontologists James Valentine, Stanley Awramik, Philip Signor, and Peter Sadler, "the single most spectacular phenomenon evident in the fossil record is the abrupt appearance and diversification of many living and extinct phyla" near the beginning of the Cambrian. Many animal body plans ranked as phyla and classes "first evolved at that time, during an interval that may have lasted no more than a few million years." Valentine and his colleagues concluded that the Cambrian explosion "was even more abrupt and extensive than previously envisioned."

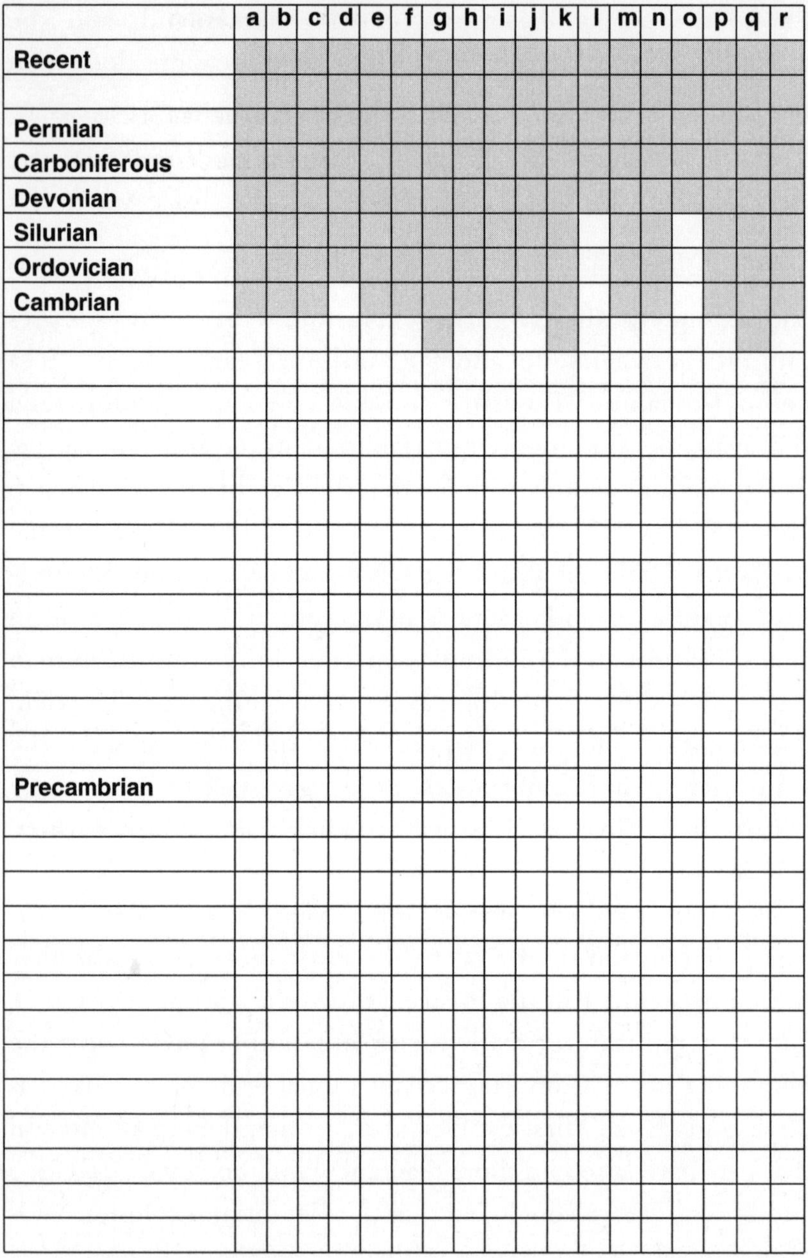

FIGURE 3-4 Actual fossil records of the major living animal phyla.

FIGURE 3-4 Actual fossil records of the major living animal phyla.

One phylum (the sponges) and possibly two others appeared just before the Cambrian; two worm phyla appeared much later, in the Carboniferous; two phyla appeared midway through the Cambrian, and one in the Ordovician. For phylum names, see the notes to this chapter at the end of the book.

The challenge to Darwin's theory

The Cambrian explosion presents a serious challenge to Darwinian evolution. The event was remarkable because it was so abrupt and extensive—that is, because it happened so quickly, geologically speaking, and because so many major groups of animals made their debut in it. But its challenge to Darwin's theory lies not so much in its abruptness (it doesn't really matter whether it lasted 5 million years or 15 million years), or in its extent (it doesn't really matter that sponges preceded it, or that some types of worms appeared later), as in the fact that phyla and classes appeared right at the start.

Darwin's theory claims that phylum- and class-level differences emerge only after a long history of divergence from lower categories such as species, genera, families and orders. Yet the Cambrian explosion is inconsistent with this picture. As evolutionary theorist Jeffrey Schwartz puts it, the major animal groups "appear in the fossil record as Athena did from the head of Zeus—full blown and raring to go."

Some biologists have described this in terms of "bottom-up" versus "top-down" evolution. Darwinian evolution is "bottom-up," referring to its prediction that lower levels in the biological hierarchy should emerge before higher ones. But the Cambrian explosion shows the opposite. In the words of Valentine and his colleagues, the Cambrian pattern "creates the impression that [animal] evolution has by and large proceeded from the 'top down'."

Clearly, the Cambrian fossil record explosion is *not* what one would expect from Darwin's theory. (Figure 3-5) Since higher levels of the biological hierarchy appear first, one could even say that the Cambrian explosion stands Darwin's tree of life on its head. If any botanical analogy were appropriate, it would be a *lawn* rather than a tree. Nevertheless, evolutionary biologists have been reluctant to abandon Darwin's theory. Many of them discount the Cambrian fossil evidence instead.

Saving Darwin's theory

There are three ways some biologists have attempted to salvage Darwin's theory in the face of the Cambrian explosion. One is to argue (as Darwin did) that the apparent absence of Precambrian ancestors is due to the fragmentary fossil record. Another is to claim that even if the record were continuous the Precambrian ancestors would not have fossilized—either because they were too small, or because they were soft-bodied. A third is to override the fossil evidence with molecular comparisons among living organisms that point to a hypothetical common ancestor hundreds of millions of years before the Cambrian.

Is the fossil record sufficiently fragmented to explain the absence of Precambrian ancestors for Cambrian animals? Most paleontologists don't think so. Enough good sedimentary rocks from the late Precambrian and Cambrian have now been found to convince paleontologists that if there had been ancestors, and they had fossilized, they would have been discovered by now. According to James Valentine and Douglas Erwin: "The sections of Cambrian rocks that we do have (and we have many) are essentially as complete as sections of equivalent time duration from similar depositional environments" in more recent rocks.

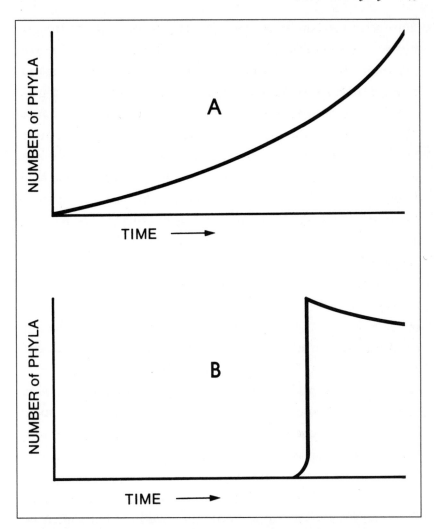

FIGURE 3-5 Evolution of the animal phyla: theory and fact.

Graphs comparing the pattern of increase in the number of animal phyla over time according to Darwin's theory and the fossil evidence. (A) In Darwin's theory, the number of animal phyla gradually increases over time. (B) The fossil record, however, shows that almost all of the animal phyla appear at about the same time in the Cambrian explosion, with the number declining slightly thereafter due to extinctions.

Yet "ancestors or intermediates" are "unknown or uncon-firmed" for any of the phyla or classes appearing in the Cambrian explosion. Valentine and Erwin conclude that the "explosion is real; it is too big to be masked by flaws in the fossil record."

Several recent surveys of the quality of the fossil record from the Cambrian to the present support this view. Although older strata are clearly not as well-preserved, on average, as younger ones, they are good enough. In February 2000, British geologists M. J. Benton, M. A. Wills, and R. Hitchin concluded: "Early parts of the fossil record are clearly incomplete, but they can be regarded as adequate to illustrate the broad patterns of the history of life."

Did the ancestors of the animal phyla fail to fossilize because they were too small, or soft-bodied? The problem with this explanation is that microfossils of tiny bacteria have been found in rocks more than three billion years old. Furthermore, the Pre-cambrian organisms found fossilized in the Australian Ediacara Hills were soft-bodied. "In the Ediacaran organisms there is no evidence for any skeletal hard parts," wrote Simon Conway Mor-ris in his 1998 book, *The Crucible of Creation*. "Ediacaran fossils look as if they were effectively soft-bodied." The same is true of many of the organisms fossilized in the Cambrian explosion. The Burgess Shale, for example, includes many fossils of completely soft-bodied animals. "These remarkable fossils," according to Conway Morris, "reveal not only their outlines but sometimes even internal organs such as the intestines or muscles."

So whatever the reason may be for the absence of ancestors, it is certainly not that they were small or soft-bodied. As geolo-gist William Schopf wrote in 1994: "There is only one source of direct evidence of the early history of life—the Precambrian fossil record; speculations made in the absence of such evidence, even by widely acclaimed evolutionists, have commonly proved groundless." One such speculation is "the long-held notion that

Precambrian organisms must have been too small or too delicate to have been preserved in geological materials." According to Schopf, this notion is "now recognized as incorrect."

The third way some evolutionary biologists have attempted to "defuse" the Cambrian explosion is by claiming that molecular evidence from living organisms points to a common ancestor of the animal phyla hundreds of millions of years before the Cambrian. In order to understand this defense of Darwin's theory—and why it doesn't work—we must turn to a relatively new discipline called "molecular phylogeny."

Molecular phylogeny

A phylogeny is the evolutionary history of a group of organisms. Until recently, phylogenies were inferred from anatomical and physiological features (such as the number of limbs, or warm-bloodedness). Since the advent of modern molecular biology, however, many phylogenies have been based on DNA and protein comparisons.

All living organisms, from bacteria to humans, contain DNA. A DNA molecule is a long chain consisting of various combinations of four subunits, abbreviated A, T, C and G; and the order of these subunits specifies the sequence of amino acids in an organism's proteins. During reproduction, the sequence of subunits is copied from one DNA molecule to another, but molecular accidents, or mutations, sometimes make the copy slightly different from the parent molecule. Therefore, organisms may have DNA molecules (and thus proteins) that differ somewhat from the DNA and proteins of their ancestors.

In 1962 biologists Emile Zuckerkandl and Linus Pauling suggested that comparisons of DNA sequences and their protein products could be used to determine how closely organisms are related. Organisms whose DNA or proteins differ by only a few subunits are presumably more closely related in evolutionary terms than those which differ by more. If mutations have accumulated steadily over time, the number of differences between organisms can serve as a "molecular clock" indicating how many years have passed since their DNA or protein was identical—that is, how long ago they shared a common ancestor. (Figure 3-6)

Much of the early work in molecular phylogeny relied on proteins, but determining protein sequences is slow work. With the development of faster techniques for determining DNA sequences, it became more common to analyze the genes coding for proteins rather than the proteins themselves. In addition to proteins and DNA, all organisms contain RNA, a close chemical relative of DNA that is involved in converting information from DNA into protein sequences. Part of this process relies on tiny particles in the cell called "ribosomes," which consist partly of ribosomal RNA, or "rRNA." Since 1980 the DNA sequences that code for rRNA have provided many of the data for molecular phylogeny.

Comparing DNA sequences is simple in theory, but complex in practice. Since an actual segment of DNA may contain thousands of subunits, lining them up to start a comparison is itself a tricky task, and different alignments can give very different results. Nevertheless, conclusions drawn from molecular comparisons have been brought to bear on the Cambrian explosion.

Molecular phylogeny and the Cambrian explosion
Did the animal phyla originate abruptly in the Cambrian, as the fossils seem to indicate, or did they slowly diverge from a common ancestor millions of years before, as Darwin's theory implies? It's not possible to analyze DNA from Cambrian fossils, but molecular biologists are able to compare protein and

DNA sequences in living species. Assuming that sequence differences among the major animal phyla are due to mutations, and that mutations accumulate at the same rate in various organisms over long periods of time, biologists use sequence differences as a "molecular clock" to estimate how long ago the phyla shared a common ancestor.

It turns out that the dates obtained by this method cover a wide range. Bruce Runnegar started the bidding in 1982 with an estimate of 900–1000 million years for the initial divergence of the animal phyla. In 1996 Russell Doolittle and his colleagues proposed a date of 670 million years, while Gregory Wray and his colleagues proposed 1200 million. In 1997 Richard Fortey and his colleagues endorsed the older date, and in 1998 Francisco Ayala and his colleagues endorsed the younger. But these two dates represent a spread of 530 million years, or as much time as has elapsed between the Cambrian explosion and the present.

	DNA Sequence
Organism 1	A T C G
Organism 2	A T C T
Organism 3	A T G T

FIGURE 3-6   Comparing DNA sequences.

All DNA molecules consist of linear sequences of four subunits, abbreviated A, T, C, and G. In the short sequence shown here, Organism 2 differs from Organism 1 in one position, while Organism 3 differs from it in two positions. If this were the only sequence being compared, Organisms 1 and 2 would be considered to have a more recent common ancestor (i.e., to be more closely related) than Organisms 1 and 3.

This "range of divergence estimates," in the opinion of American geneticist Kenneth Halanych, testifies "against the ability to date such ancient events" using molecular methods.

Obviously, 670 million years comes closer to fitting the fossil record than 1200 million. For some scientists, the choice between the two comes down to a choice between molecular and paleontological evidence. In 1998 molecular evolutionists Lindell Bromham, Andrew Rambault, Richard Fortey, Alan Cooper, and David Penny relied on molecular data "to confidently reject the Cambrian explosion hypothesis, which rests on a literal interpretation of the fossil record." In 1999, however, paleontologists James Valentine, David Jablonski, and Douglas Erwin wrote that "the accuracy of molecular clocks is still problematical, at least for phylum divergences," since the estimates vary by hundreds of millions of years "depending on the techniques or molecules used." Valentine and his colleagues consider the fossil record to be the primary evidence, and maintain that the molecular data "do not muffle the [Cambrian] explosion, which continues to stand out as a major feature" in animal evolution.

So the Cambrian explosion remains a paradox. The fossil evidence shows that the major animal phyla and classes appeared right at the start, contradicting a major tenet of Darwin's theory. Molecular phylogeny has not resolved the paradox, because the dates inferred from it vary over such a wide range.

The failure of molecular phylogeny to resolve the paradox now appears to be part of a larger problem. Since the early 1970s, evolutionary biologists have been hoping that sequence comparisons would overcome many of the difficulties arising from more traditional approaches, and would enable them to construct a "universal tree of life" based on molecules alone. Recent discoveries, however, have dashed that hope.

The growing problem in molecular phylogeny

Modern versions of the Darwinian tree of life are called "phylogenetic trees." In a typical phylogenetic tree, the "root" is the common ancestor of all the other organisms in the tree. The lower branches represent lineages that diverged relatively early, while the upper branches diverged later. The tips of the branches are actual species. Wherever two branches diverge, the branch-point indicates the hypothetical common ancestor of the two branching lineages. Many phylogenetic trees are drawn so that the lengths of the branches are proportional to sequence differences, which are often assumed to indicate how much time has elapsed since lineages diverged. (Figure 3-7)

It is important to remember that the only actual data in a phylogenetic tree (with rare exceptions) come from living organisms, which are the tips of the branches. Everything else about a phylogenetic tree is hypothetical. The arrangement of the tips, the branches and branch-points, and the root itself are all based on methodological assumptions and sequence comparisons.

Ideally, phylogenetic trees should be approximately the same regardless of which molecules are chosen for comparison. Indeed, there has been a general expectation among evolutionary biologists that the more molecules they include in a phylogenetic analysis, the more reliable their results are likely to be.

But the expectation that more data would help matters "began to crumble a decade ago," wrote University of California molecular biologists James Lake, Ravi Jain, and Maria Rivera in 1999, "when scientists started analyzing a variety of genes from different organisms and found that their relationships to each other contradicted the evolutionary tree of life derived from rRNA analysis alone." According to French biologists Hervé Philippe and Patrick Forterre: "With more and more sequences available, it turned out that most protein phylogenies contradict each other as well as the rRNA tree."

In other words, different molecules lead to very different phylogenetic trees. According to University of Illinois biologist Carl Woese, an early pioneer in constructing rRNA-based phylogenetic trees: "No consistent organismal phylogeny has emerged from the many individual protein phylogenies so far produced. Phylogenetic incongruities can be seen everywhere in the universal tree, from its root to the major branchings within and among the various [groups] to the makeup of the primary groupings themselves."

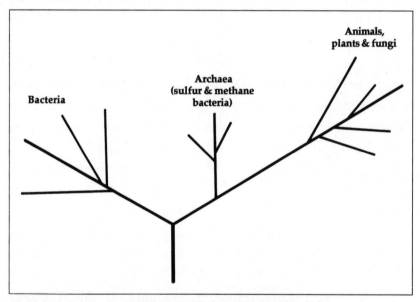

FIGURE 3-7 **A molecular phylogenetic tree, circa 1990.**

A tree based on rRNA genes, showing the presumed evolutionary relationships among the kingdoms of life. The root represents the universal common ancestor; lower branches represent lineages that presumably diverged before the upper branches; and the branch-points represent the hypothetical immediate common ancestors of the lineages that diverge from them.

Woese dealt mainly with discrepancies at the level of the major kingdoms of life, but (as he indicated) the problems extend even to smaller branches, including animal phylogenies. "Clarification of the phylogenetic relationships of the major animal phyla has been an elusive problem," wrote biologist Michael Lynch in 1999, "with analyses based on different genes and even different analyses based on the same genes yielding a diversity of phylogenetic trees." Even when different molecules can be combined to give a single tree, the result is often bizarre: A 1996 study using 88 protein sequences grouped rabbits with primates instead of rodents; a 1998 analysis of 13 genes in 19 animal species placed sea urchins among the chordates; and another 1998 study based on 12 proteins put cows closer to whales than to horses.

Inconsistencies among trees based on different molecules, and the bizarre trees that result from some molecular analyses, have now plunged molecular phylogeny into a crisis.

Uprooting the tree of life

Some molecular biologists believe that the problem is methodological. According to Forterre and Philippe, some sequences evolve too rapidly to preserve a "phylogenetic signal" over long periods of time. They claim that by limiting themselves to sequences they believe evolved slowly, they can produce a consistent universal tree. The problem is that their analysis points to a cell with a nucleus as the universal common ancestor. Since bacteria (which do not have nuclei) are simpler than cells with nuclei, Darwinists have traditionally believed that the latter evolved from the former. In other words, from the standpoint of Darwinian evolution the phylogenetic tree proposed by Forterre and Philippe is rooted in the wrong place.

Other biologists think the problem is not just methodological. For example, Woese maintains that the incongruities "are sufficiently frequent and statistically solid that they can neither be overlooked nor trivially dismissed on methodological grounds." According to Woese, "it is time to question underlying assumptions."

Woese recommends abandoning the idea that the universal common ancestor is a living organism. "The universal ancestor is not an entity, not a thing," wrote Woese in 1998, "it is a process." As Woese conceives it, that process did not involve organisms "in any conventional sense," but an interchange of genetic material in a complex primordial soup. He concludes: "The universal phylogenetic tree, therefore, is not an organismal tree at its base." But if the universal common ancestor was not an organism, then does it make sense to call it an "ancestor"? If the primordial soup is our ancestor, so is the periodic table of the elements, or the planet Earth. Once the notion of organism is discarded, the word "ancestor" loses its biological meaning.

Another solution to the problem has been proposed by Dalhousie University biologist W. Ford Doolittle. Maybe molecular phylogeneticists "have failed to find the 'true tree'," wrote Doolittle in 1999, "not because their methods are inadequate or because they have chosen the wrong genes, but because the history of life cannot properly be represented as a tree." According to Doolittle, the discrepancies in molecular phylogenies are due largely to "lateral gene transfer." Microbiologists know that bacteria can exchange genes, and Doolittle proposes that gene exchange among bacteria and primitive cells with nuclei could account for many of the discrepancies we now see in molecular phylogenies. But then the early history of life would not have resembled a branching tree, but a tangled thicket. (Figure 3-8)

According to Doolittle: "Perhaps it would be easier, and in the long run more productive, to abandon the attempt to force the data that Zuckerkandl and Pauling stimulated biologists to collect into the mold provided by Darwin." In a February 2000 article in *Scientific American* entitled "Uprooting the Tree of Life," Doolittle concluded: "Now new hypotheses, having final forms we cannot yet guess, are called for."

So the branching-tree pattern of evolution is inconsistent with major features of the fossil and molecular evidence. The Cambrian explosion demonstrates that the highest categories of animals appeared first, thus turning Darwin's tree of life upside down. The molecular evidence, far from saving it, uproots it entirely. Yet the tree of life still dominates the iconography of evolution, because Darwinists have declared it to be a fact.

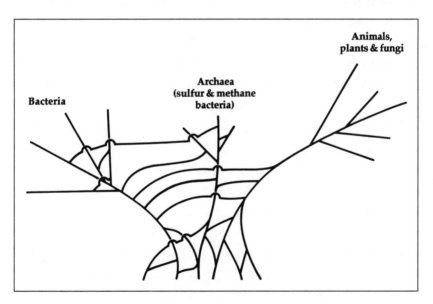

FIGURE 3-8 **The molecular thicket of life (as of 2000).**

This diagram attempts to take into account both the absence of a single universal common ancestor and some of the lateral gene transfer that has supposedly occurred throughout the history of life. The resulting pattern is less like a tree than a tangled thicket.

The fact of evolution

For many years, the California Academy of Sciences in San Francisco has proudly featured a museum exhibit about evolution. As parents, teachers, and schoolchildren wander through the exhibit, their attention is occasionally drawn to magnifying glasses mounted over tiny fossils in the display cases. Visitors reaching the end of the exhibit are treated to the "Hard Facts Wall," which shows a phylogenetic tree of the major animal phyla. The various branch points in the tree—indicating supposed common ancestors—are decorated with magnifying glasses like those elsewhere in the exhibit. But as tired visitors pass by the Hard Facts Wall on their way to the exit, most of them miss the fact that *these* magnifying glasses have nothing under them. There are no "hard facts" there to see.

Maybe hard facts seemed superfluous to the exhibit's creators, because people have become conditioned to thinking that the Darwinian tree of life is *itself* a fact. According to the same 1998 National Academy of Sciences booklet mentioned in the previous chapters: "Scientists most often use the word 'fact' to describe an observation. But scientists can also use fact to mean something that has been tested or observed so many times that there is no longer a compelling reason to keep testing or looking for examples. The occurrence of evolution in this sense is a fact. Scientists no longer question whether descent with modification occurred because the evidence supporting the idea is so strong."

The booklet is not talking about descent with modification *within* a species, because no one ever questioned that anyway. It is claiming that descent with modification of all organisms from common ancestors is a fact, and it lists "several compelling lines of evidence that demonstrate [this] beyond a reasonable doubt."

These lines of evidence include the fossil record, common anatomical structures, the geographical distribution of species, similarities during embryo development, and DNA sequences.

The authors of a 1999 booklet also published by the National Academy go into more detail on the first of these: "The fossil record thus provides consistent evidence of systematic change through time—of descent with modification." Yet there is no mention at all of the Cambrian explosion, or of the paradox it presents for Darwinian evolution, though both have been well known for over a decade. The Cambrian explosion even made the cover of *Time* magazine in 1995.

Regarding molecular phylogeny, the 1999 booklet continues: "As the ability to sequence… DNA has improved, it has also become possible to use genes to reconstruct the evolutionary history of organisms." The booklet concludes: "The evidence for evolution from molecular biology is overwhelming and is growing quickly." What the booklet doesn't mention, however, is that this growing evidence *uproots* the standard evolutionary history of life.

One might be tempted to excuse the booklet's authors for ignoring the last three years of published articles in molecular phylogeny, on the grounds that they cannot be expected to keep up with all the research. But they also ignored the fossil evidence from the Cambrian explosion, and (as we saw in the last chapter) the evidence that the Miller-Urey experiment failed to simulate primitive earth conditions. These writers purport to be representing the nation's premier science organization, yet even ordinary scientists are expected to keep up with research in their field—especially if they are going to write authoritative-sounding booklets about it.

Since booklets published by the National Academy of Sciences ignore the fossil and molecular evidence and call evolution a "fact," perhaps it is not surprising to find biology textbooks doing the same. "Descent with modification from common ancestors is a scientific *fact*, that is, a hypothesis so well supported by evidence that we take it to be true" according to Douglas Futuyma's 1998 college textbook, *Evolutionary Biology*. "The *theory* of evolution, on the other hand, is a complex body of statements, well supported but still incomplete, about the causes of evolution." (emphasis in original) Although Futuyma's book subsequently discusses the Cambrian explosion, its emphasis is on explaining it away rather than dealing candidly with its challenge to Darwin's theory.

Distinguishing between fact and theory—and insulating universal common descent from criticism by placing it on the "fact" side of the divide—is typical of other biology textbooks, as well. For example, the 1999 edition of *Biology*, by Neil Campbell, Jane Reece, and Lawrence Mitchell—probably the most widely used introductory college biology textbook in the United States—explains that "Darwinism has a dual meaning." The first is the historical fact that "all organisms [are] related through descent from some unknown prototype that lived in the remote past," so that "the history of life is like a tree." The second is "Darwin's theory of natural selection—the mechanism Darwin proposed to explain the historical facts" included in the first meaning.

Anyone reading these books without knowing better would get the impression that the evidence for the Darwinian tree of life is overwhelming, and that no scientist would think of doubting universal common descent. Yet Harry Whittington, the renowned paleontologist whose work first revealed the extent of the Cambrian explosion in the Burgess Shale, did not hesitate to doubt it. Whittington wrote in 1985: "I look skeptically upon diagrams that show the branching diversity of animal life through time, and come down at the base to a single kind of animal.... Animals may have originated more than once, in different places and at different times."

And Whittington did not even know about the recent evidence from molecular phylogeny. Biologist Malcolm Gordon, who does know about it, wrote in 1999 that "life appears to have had many origins. The base of the universal tree of life appears not to have been a single root." Gordon concluded: "The traditional version of the theory of common descent apparently does not apply to kingdoms... [or] phyla, and possibly also not to many classes within the phyla."

Clearly, qualified biologists can and do question the Darwinian tree of life. Nevertheless, some influential writers continue to insist that evolution—in the sense of descent with modification from common ancestors—is a "fact." But unless they are referring only to what happens within a species, this is about as far from the truth as one can get. At the level of kingdoms, phyla, and classes, descent with modification from common ancestors is obviously *not* an observed fact. To judge from the fossil and molecular evidence, it's not even a well-supported theory.

So why does the tree of life continue to be such a popular icon of evolution? The best way for biology students to find out might be to ask those who continue to use it. But their question may not be warmly welcomed, at least in the United States. In 1999, a Chinese paleontologist who is an acknowledged expert on Cambrian fossils visited the United States to lecture on several university campuses. I attended one lecture in which he pointed out that the "top-down" pattern of the Cambrian explosion contradicts Darwin's theory of evolution. Afterwards, scientists in the audience asked him many questions about specific fossils, but they completely avoided the topic of Darwinian evolution. When our Chinese visitor later asked me why, I told him that perhaps they were just being polite to their visitor, because criticizing Darwinism is unpopular with American scientists. At that he laughed, and said: "In China we can criticize Darwin, but not the government; in America, you can criticize the government, but not Darwin."

CHAPTER 4

Homology in Vertebrate Limbs

Biologists since Aristotle have noticed that very different organisms may share remarkable similarities. One kind of similarity is functional: Butterflies have wings for flying, and so do bats, but the two animals are constructed very differently. Another kind of similarity is structural: The pattern of bones in a bat's wing is similar to that in a porpoise's flipper, though the wing is used for flying and the flipper is used for swimming.

In the 1840s British anatomist Richard Owen called the first kind of similarity "analogy," and the second kind "homology." At the time, the distinction served principally as an aid in biological classification: Analogy suggests independent adaptations to external conditions, while homology suggests deeper structural affinities. The latter was considered a more reliable guide in grouping organisms together in families, orders, classes and phyla.

The classic examples of homologous structures are the forelimbs of vertebrates (animals with backbones). Although a bat has wings for flying, a porpoise has flippers for swimming, a horse has legs for running, and a human has hands for grasping, the bone patterns in their forelimbs are similar. (Figure 4-1) Such skeletal similarities, along with other internal affinities such as warm-bloodedness and milk production, justify classifying all these creatures as mammals despite their external differences.

Like other pre-Darwinian biologists, Owen considered homologous features to be derived from a common "archetype." An "archetype," however, could be understood in various ways: a disembodied Platonic idea, a plan in the mind of the Creator, an Aristotelian form inherent in the structure of nature, or a prototypical organism, among others. Both Owen and Darwin

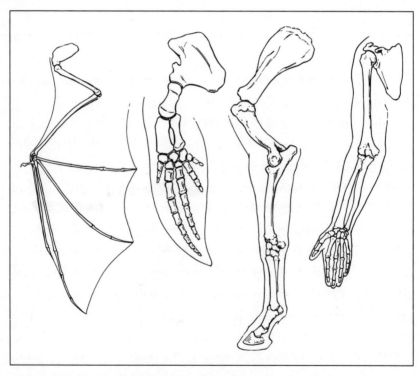

FIGURE 4-1   **Homology in vertebrate limbs.**

Forelimbs of (a) bat, (b) porpoise, (c) horse, and (d) human, showing bones considered to be homologous.

regarded the archetype as a prototypical organism, but Owen was not an evolutionist. While Owen regarded organisms as constructed on a common plan, Darwin regarded them as descended from a common ancestor.

In *The Origin of Species* Darwin argued that the best explanation for homology is descent with modification. "If we suppose that an early progenitor—the archetype as it may be called—of all mammals, birds and reptiles, had its limbs constructed on the existing pattern," then "the similar framework of bones in the hand of a man, wing of a bat, fin of the porpoise, and leg of the horse... at once explain themselves on the theory of descent with slow and slight modifications." Darwin considered homology important evidence for evolution, listing it among the facts which "proclaim so plainly, that the innumerable species, genera and families, with which this world is peopled, are all descended, each within its own class or group, from common parents."

The link between homology and common descent was so central to Darwin's theory that his followers actually *re-defined* homology to mean features inherited from a common ancestor. Even after homology was re-defined, however, the Darwinian account remained incomplete without a mechanism to explain why homologous features were so similar in such different organisms. When neo-Darwinism arose in the 1930s and 1940s, it seemed to have a solution to this problem: Homologous features were attributed to similar genes inherited from a common ancestor.

Modern Darwinists continue to use homology as evidence for their theory. In fact, next to the Darwinian tree of life, homology in vertebrate limbs is probably the most common icon of evolution in biology textbooks. But the icon conceals two serious problems: First, if homology is *defined* as similarity due to common descent, then it is circular reasoning to use it as *evidence* for common descent. Second, biologists have known for decades that homologous features are not due to similar genes, so the mechanism that produces them remains unknown.

Re-defining homology

For Darwin, homologies were similar structures explained by common ancestry. But some similar structures are *not* acquired through common ancestry. For example, the structure of an octopus eye is remarkably similar to the structure of a human eye, yet biologists do not think that the common ancestor of octopuses and humans possessed such an eye. To ensure that only structures inherited from a common ancestor would be called homologous, Darwin's followers redefined homology to mean similarity due to common ancestry.

So before Darwin (and for Darwin himself), the definition of homology was similarity of structure and position (as in the bone patterns of vertebrate limbs). But similarity of structure and position did not explain the origin of homology, so an explanation had to be provided. For pre-Darwinian biologists, the explanation was derivation from an original pattern, or archetype. Darwin identified "derivation" with biological evolution, and "archetype" with a common ancestor.

But for twentieth-century neo-Darwinists, common ancestry is the *definition* of homology as well as its *explanation*. According to Ernst Mayr, one of the principal architects of neo-Darwinism: "After 1859 there has been only one definition of homologous that makes biological sense.... Attributes of two organisms are homologous when they are derived from an equivalent characteristic of the common ancestor."

In other words, with Charles Darwin evolution was a theory, and homology was evidence for it. With Darwin's followers, evolution is assumed to be independently established, and homology is its result. The problem is that now homology cannot be used as evidence for evolution except by reasoning in a circle.

Homology and circular reasoning

Consider the example of bone patterns in forelimbs (Figure 4-1), which Darwin regarded as evidence for the common ancestry of the vertebrates. A neo-Darwinist who wants to determine whether vertebrate forelimbs are homologous must first determine whether they are derived from a common ancestor. In other words, there must be evidence for common ancestry before limbs can be called homologous. But then to turn around and argue that homologous limbs point to common ancestry is a vicious circle: Common ancestry demonstrates homology which demonstrates common ancestry. (Figure 4-2)

This circularity has been noticed and criticized by many biologists and philosophers. In 1945 J. H. Woodger wrote that the new definition was "putting the cart before the horse." Alan Boyden pointed out in 1947 that neo-Darwinian homology requires "that we first *know* the ancestry and then decide that the corresponding organs or parts" are homologous. "*As though we could know the ancestry without the essential similarities to guide us!*" (emphasis in original) When neo-Darwinian paleontologist George Gaylord Simpson tried to use homology-as-common-ancestry to infer evolutionary relationships, biologists Robert Sokal and Peter Sneath criticized him for "the circularity of reasoning" inherent in his procedure.

Neo-Darwinian philosophers rose to the defense. In 1966 Michael Ghiselin pointed out that the neo-Darwinian definition is not circular because homology is not defined in terms of itself. But this did not solve the problem, because although the definition is not circular, the reasoning based on it is. The fol-

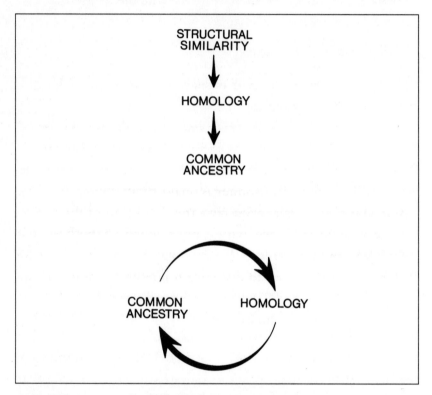

FIGURE 4-2 homology and circular reasoning.

(Top) Darwin, like his predecessors, inferred homology from structural similarity, then inferred common ancestry from homology. (Bottom) in the circular reasoning employed by some modern neo-Darwinists, homology is inferred from common ancestry, then turned around and used as evidence for common ancestry.

lowing year, David Hull argued that the reasoning is not circular, but merely an example of the scientific "method of successive approximation" (or what German biologist Willi Hennig called the "method of reciprocal illumination"). According to Hull, evolutionary biologists start by assuming a particular hypothesis of descent, then they use similarities to refine the hypothesis. But the method—which critics at the time derided as "groping"—works, if it works at all, only by assuming the truth of common ancestry. If the question is whether Darwin's theory is true in the first place, then Hull's method of successive approximation is just another circular argument.

The controversy has raged ever since. Neo-Darwinists defend their notion of homology as common ancestry, while critics object that it confuses definition with explanation and leads to circular reasoning. Philosopher Ronald Brady wrote in 1985: "By making our explanation into the definition of the condition to be explained, we express not scientific hypothesis but belief. We are so convinced that our explanation is true that we no longer see any need to distinguish it from the situation we were trying to explain. Dogmatic endeavors of this kind must eventually leave the realm of science."

Breaking the circle

There seem to be only three ways to avoid the circular reasoning brought on by defining and explaining homology in terms of common ancestry. One way is to embrace the neo-Darwinian definition but give up trying to infer common descent from it—in other words, to acknowledge that homology no longer provides evidence for evolution. "Common ancestry is all there is to homology," wrote evolutionary biologist David Wake in 1999;

thus "homology is the anticipated and expected consequence of evolution. Homology is not evidence of evolution."

A second way is to retain the pre-Darwinian definition of homology as structural similarity, but acknowledge that this re-opens the question of whether descent with modification is the best explanation for it. Recent advocates of this position are hard to find, because among biologists in the United States it is extremely unpopular (and professionally risky) to question whether Darwinian evolution is the best explanation.

The third (and currently most popular) way to deal with the problem is to define homology in terms of common ancestry and then seek evidence for descent with modification that is inde-pendent of homology. Such evidence may come from pattern (DNA sequence comparisons or the fossil record) or process (developmental pathways and developmental genetics). The first two begin by assuming common ancestry, and then attempt to infer the most likely pattern of ancestor-descendant relationships. The second two attempt to identify the processes that would account for similarity due to common ancestry.

Evidence from DNA sequences

As we saw in the previous chapter, molecular phylogenies are constructed by comparing DNA sequences (or their protein products) in different organisms. Since DNA sequences are copied directly from other DNA sequences through the process of replication, molecular phylogeneticists assume that sequence similarities are more likely to indicate an ancestor-descendant relationship than morphological similarities, which are produced by a complex series of events in the embryo rather than inherited directly from parents.

Unfortunately, molecular sequence comparisons face as many difficulties as morphological comparisons. First, in molecular phylogeny the meaning of "homology" is no less problematic. As molecular biologist David Hillis wrote in 1994, "the word homology is now used in molecular biology to describe every-thing from simple similarity (whatever its cause) to common ancestry (no matter how dissimilar the structures)." Thus "molecular biologists may have done more to confound the meaning of the term homology than have any other group of scientists."

Second, identifying homologous sequences is as difficult as identifying homologous organs. According to Hillis: "Some pro-ponents of molecular techniques have claimed that molecular biology 'solves the problem of homology'... [but] the difficul-ties of assigning homology to molecules parallel many of the difficulties of assigning homology to morphological structures."

Finally, molecular homology generates at least as many con-flicting results as the more traditional approach. "Congruence between molecular phylogenies," wrote British biologists Colin Patterson, David Williams and Christopher Humphries in 1993, "is as elusive as it is in morphology." But when molecular phylogenies conflict, the only way to choose among them is to have independent knowledge of common ancestry, and this leads right back into the very circular reasoning that molecular com-parisons were supposed to avoid.

The fossil record

How about the fossil record? Some biologists have argued the best way to determine evolutionary relationships would be to trace the similarities in two or more organisms back through an unbroken chain of fossil organisms to their common ancestor. Unfortunately, comparing fossils is no more straightforward than comparing live specimens. As Sokal and Sneath pointed out in 1963: "Even when fossil evidence is available, this evidence itself must first be interpreted" by comparing similar features. Any attempt to infer evolutionary relationships among fossils based on homology-as-common-ancestry "soon leads to a tangle of circular arguments from which there is no escape."

In fact, inferring evolutionary relationships from the fossil record is *more* difficult than inferring them from live specimens, because the record is fragmentary and because fossils do not preserve all relevant features. As biologist Bruce Young wrote in 1993: "If anything, fossils are of less value in establishing homologues since they normally include far fewer characters" than living organisms.

But even if the fossil record were complete, and it preserved all the desired characters, it would not establish that homology is due to common ancestry. This problem was inadvertently illustrated by biologist Tim Berra in a 1990 book defending Darwinian evolution against creationist critics. Berra compared the fossil record to a series of automobile models: "If you compare a 1953 and a 1954 Corvette, side by side, then a 1954 and a 1955 model, and so on, the descent with modification is overwhelmingly obvious. This is what [paleontologists] do with fossils, *and the evidence is so solid and comprehensive that it cannot be denied by reasonable people.*" (emphasis in the original)

But Berra's analogy actually spotlights the problem of using a sequence of similarities as evidence for Darwin's theory. We all know that automobiles are manufactured according to archetypes (in this case, plans drawn up by engineers), so it is clear that there can be other explanations for a sequence of similarities besides descent with modification. In fact, most pre-Darwinian biologists would have explained such sequences by something akin to automobile manufacturing—that is, creation by design. So although Berra believed he was defending Darwinian evolution against creationist explanations, he unwittingly showed that the fossil evidence is compatible with either. Law professor (and critic of Darwinism) Phillip E. Johnson dubbed this "Berra's Blunder." (Figure 4-3)

Berra's Blunder demonstrates that a mere succession of similar forms does not furnish its own explanation. Something more is needed—a mechanism. In the case of Corvettes, the mechanism (human manufacturing) can be directly observed; but in a succession of fossils, it cannot. This is where Darwin's theory comes in. For Darwin, the mechanism is descent with modification. But "descent" and "modification" are merely words, unless they can be tied to actual biological processes.

Darwin realized this. He wrote in *The Origin of Species* that a naturalist reflecting on the geological evidence "might come to the conclusion that species had not been independently created, but had descended, like varieties, from other species. Nevertheless, such a conclusion, even if well founded, would be unsatisfactory, until it could be shown how the innumerable species inhabiting this world have been modified." Darwin concluded: "It is, therefore, of the highest importance to gain a clear insight into the means of modification."

Of course, the means of modification in Darwin's theory is natural selection. But the means of *descent* remained elusive. In the ordinary process of reproduction, like always produces like. Can natural selection alter the process, so that like sometimes produces not-so-like? Darwin didn't know enough about embryo development to answer the question. Without knowing the mechanisms that make embryos similar, it is mere speculation to say that those unknown mechanisms can be modified by natural selection.

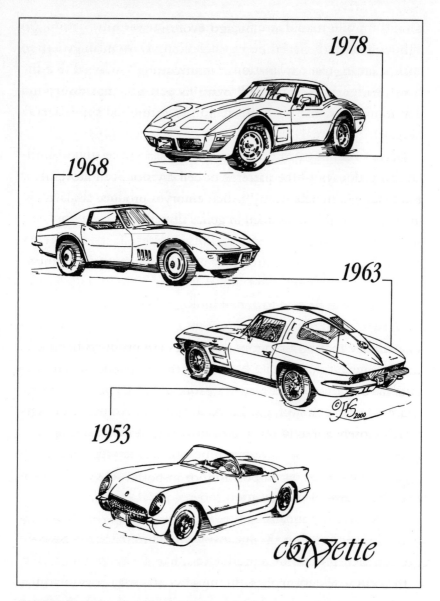

FIGURE 4-3   Berra's Blunder.

Berra used four models of Corvette automobiles to illustrate descent with modification. Shown here from bottom to top: 1953, 1963, 1968, and 1978 models.

In 1982 University of Chicago evolutionary biologist Leigh Van Valen wrote that the key to explaining homology lies in understanding the "continuity of information." An embryo contains information, inherited from its parents, that directs its development. Until we understand the nature of that information, we cannot understand how it might be modified.

Developmental information could be in the form of "developmental pathways"—the patterns of cell division, cell movement, and tissue differentiation by which embryos produce adult structures. Or it could be encoded in genes that affect the development of the embryo. But neither developmental pathways nor developmental genetics has solved the problem of what causes homology.

Evidence from developmental pathways

The theory that homologous structures are products of similar developmental pathways does not fit the evidence, and biologists have known this for over a century. "It is a familiar fact," said American embryologist Edmund Wilson in 1894, "that parts which closely agree in the adult, and are undoubtedly homologous, often differ widely in larval or embryonic origin either in mode of formation or in position, or in both." More than sixty years later, after reviewing the embryological evidence that had been amassed since Wilson's time, British biologist Gavin de Beer agreed: "The fact is that correspondence between homologous structures cannot be pressed back to similarity of position of the cells in the embryo, or of the parts of the egg out of which the structures are ultimately composed, or of developmental mechanisms by which they are formed."

De Beer's assessment is still accurate. It is "the rule rather than the exception," developmental biologist Pere Alberch wrote in

1985, that "homologous structures form from distinctly dissimilar initial states." Evolutionary developmental biologist Rudolf Raff, who studies two species of sea urchin that develop by radically different pathways into almost identical adult forms, restated the problem in 1999: "Homologous features in two related organisms should arise by similar developmental processes.... [but] features that we regard as homologous from morphological and phylogenetic criteria can arise in different ways in development."

The lack of correspondence between homology and developmental pathways is true not only in general, but also in the particular case of vertebrate limbs. The classic examples of this problem are salamanders. In most vertebrate limbs, development of the digits proceeds from posterior to anterior—that is, in the tail-to-head direction. This accurately describes frogs, but their fellow amphibians, salamanders, do it differently. In salamanders, development of the digits proceeds in the opposite direction, from head to tail. The difference is so striking that some biologists have argued that the evolutionary history of salamanders must have been different from all other vertebrates, including frogs.

There are other anomalies, as well. Skeletal patterns in vertebrate limbs initially form as cartilage, which later turn into bone. If the development of vertebrate limbs reflected their origin in a common ancestor, one might expect to see a common ancestral cartilage pattern early in vertebrate limb development. But this is not the case. Cartilage patterns correspond to the form of the adult limb from the beginning, not only in salamanders, but also in frogs, chicks and mice. According to British zoologists Richard Hinchliffe and P. J. Griffiths, the idea that vertebrate limbs develop from a common ancestral pattern in the embryo "has arisen because investigators have superimposed their pre-conceptions" on the evidence.

So homologous features, even in vertebrate limbs, are not produced by similar developmental pathways. How about similar genes?

Evidence from developmental genetics

According to neo-Darwinian theory, the information Van Valen described is contained in DNA sequences, or genes. Genes carry information from one generation to the next, and according to theory direct the development of the embryo. Therefore, the neo-Darwinian explanation for homologous features is that they are programmed by similar genes inherited from a common ancestor. If it could be shown that homologous structures in two different organisms are produced by similar genes, and that homologous structures are not produced by different genes, then we would have evidence for the "continuity of information" that Van Valen wrote about.

But this is not the case, and biologists have known it for decades. In 1971 Gavin de Beer wrote: "Because homology implies community of descent from... a common ancestor it might be thought that genetics would provide the key to the problem of homology. This is where the worst shock of all is encountered... [because] characters controlled by identical genes are not necessarily homologous... [and] homologous structures need not be controlled by identical genes." De Beer concluded that "the inheritance of homologous structures from a common ancestor... cannot be ascribed to identity of genes."

To illustrate his point that homologous structures can arise from different genes, de Beer cited only one experiment (involving eye development in fruit flies), but other examples have been found since then. One involves segment formation in insects. Fruit fly embryos require the gene *even-skipped* for the proper development of body segments; but other insects, such as locusts and wasps, form segments without using this gene. Since all insect segments are considered homologous (whether defined in terms of structural similarity or common ancestry), this shows that homologous features need not be controlled by identical genes. Another example is *Sex-lethal*, a gene that is required for sex-determination in fruit flies but not in other insects, which produce males and females without it.

The opposite situation—non-homologous structures arising from identical genes—is both more striking and more common. Geneticists have found that many of the genes required for proper development in fruit flies are similar to genes in mice, sea urchins, and even worms. In fact, gene transplant experiments have shown that developmental genes from mice (and humans) can functionally replace their counterparts in flies. If genes control structure, and the developmental genes of mice and flies are so similar, why doesn't a mouse embryo develop into a fly, or a fly embryo into a mouse?

The lack of correspondence between genes and structures is true not only for entire organisms, but also for limbs. One developmental gene shared by several different types of animals is *Distal-less*, so named because a mutation in it blocks limb development in fruit flies ("distal" refers to structures away from the main part of the body). A gene with a very similar DNA sequence has been found in mice; in fact, genes similar to *Distal-less* have been found in sea urchins, spiny worms (members of the same phylum as earthworms), and velvet worms (another phylum entirely). (Figure 4-4)

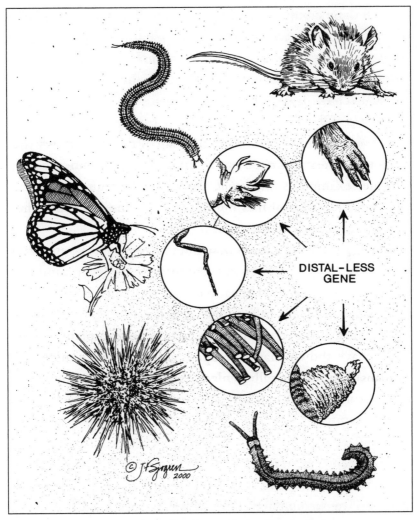

FIGURE 4-4 **A similar gene in non-homologous limbs.**

The gene *Distal-less* is involved in the development of appendages in all five of these animals, yet the appendages are not homologous either by similar structure or by common ancestry. The animals, each in a different phylum, are (counterclockwise from top): mouse; spiny worm; butterfly; sea urchin (its limbs are tube feet underneath its body); and velvet worm.

In all these animals, *Distal-less* is involved in the development of appendages, yet the appendages of these five groups of animals are not structurally or evolutionarily homologous. "These similarities are puzzling," noted the biologists who reported them in 1997, because the "appendages have such vastly different anatomies and evolutionary histories." In 1999 Gregory Wray found "surprising" the association between *Distal-less* and "what are superficially similar, but non-homologous structures." He concluded: "This association between a regulatory gene and several non-homologous structures seems to be the rule rather than the exception."

Not only *Distal-less* but also the entire network of genes involved in limb development has been found to be similar in insects and vertebrates. Clifford Tabin, Sean Carroll, and Grace Panganiban, who described these networks in 1999, noted that "there has been no continuity of any structure from which the insect and vertebrate appendages could be derived, i.e., they are not homologous structures. However, there is abundant evidence for continuity in the genetic information" involved in their development.

Evolutionary biologists argue that the striking similarity of developmental genes in such a wide variety of animal phyla points to their common ancestry. And so it might. But then the problems we encountered above with molecular phylogenies surface again, while the problem of explaining how homologous structures arise remains unsolved.

The conclusion is clear: Whether or not homology is due to descent with modification, the specific mechanism responsible for producing it remains unknown. In 1971 Gavin de Beer wrote: "What mechanism can it be that results in the production of homologous organs, the same 'patterns', in spite of their *not* being controlled by the same genes? I asked this question in 1938, and it has not been answered." Today, more than sixty years after it was first asked, de Beer's question still has not been answered.

Vertebrate limbs as evidence for evolution?

How do vertebrate limbs provide evidence for Darwinian evolution? If continuity of information does not come from genes or developmental pathways, how do we know that it comes from descent with modification? Is it even possible to infer common ancestry from homology? If we attempt to settle the issue simply by *defining* homology as common ancestry, how can we then use homology as evidence for evolution? These are legitimate scientific questions, but biology students will probably not find them in their textbooks.

Almost every biology textbook uses vertebrate limbs to illustrate homology, and claims that homology is evidence for common ancestry. But most of those textbooks also *define* homology in terms of common ancestry. They thereby fall into the same vicious circle that biologists and philosophers have been criticizing for over half a century.

For example, the 1999 edition of Teresa and Gerald Audesirk's *Biology: Life on Earth* explains that "internally similar structures are called homologous structures, meaning that they have the same evolutionary origin," and on the very same page states that homologous structures "provide evidence of relatedness in organisms." Along the same lines, the most recent edition of Sylvia Mader's *Biology* declares: "Structures that are similar because they were inherited from a common ancestor are called homologous structures," and on the same page claims: "This unity of plan is evidence of a common ancestor."

According to the 1999 edition of Peter Raven and George Johnson's *Biology*, homology refers to "structures with different appearances and functions that all derived from the same body part in a common ancestor," yet the book also claims that homology is "evidence of evolutionary relatedness." And the 1999 edition of Campbell, Reece, and Mitchell's *Biology* contains the following: "Similarity in characteristics resulting from common ancestry is known as homology, and such anatomical signs of evolution are called homologous structures. Comparative anatomy is consistent with all other evidence in testifying [to] evolution."

These textbooks give students no hint of the continuing controversy over homology. Instead, they give the impression that it is scientific to define homology in terms of common ancestry and then turn around and claim it as evidence for common ancestry. Such circular reasoning lulls students into sloppy and uncritical thinking. This is a problem not just for science, but for our society as a whole. A democracy needs well-educated citizens who can spot faulty arguments and think for themselves, not docile masses who swallow what they are fed by authority figures.

Critical thinking in action

Faced with the circular reasoning prevalent in most biology textbooks, students might do well to ask more questions in class. According to Henry Gee, Chief Science Writer for the prestigious journal, *Nature*, "nobody should be afraid to ask a silly question." In science, Gee writes, "statements from authorities in a field should be as subject to scrutiny as those emanating from the most humble sources, even a beginning student."

What would happen if a beginning student were to ask some appropriately respectful questions about homology? One might imagine the following exchange between an inquisitive student and a biology teacher:

Teacher: OK, let's start today's lesson with a quick review. Yesterday I talked about homology. Homologous features, such as the vertebrate limbs shown in your textbook, provide us with some of our best evidence that living things have evolved from common ancestors.

Student (raising hand): I know you went over this yesterday, but I'm still confused. How do we know whether features are homologous?

Teacher: Well, if you look at vertebrate limbs, you can see that even though they're adapted to perform different functions their bone patterns are structurally similar.

Student: But you told us yesterday that even though an octopus eye is structurally similar to a human eye, the two are *not* homologous.

Teacher: That's correct. Octopus and human eyes are not homologous because their common ancestor did not have such an eye.

Student: So regardless of similarity, features are not homologous unless they are inherited from a common ancestor?

Teacher: Yes, now you're catching on.

Student (looking puzzled): Well, actually, I'm still confused. You say homologous features provide some of our best evidence for common ancestry. But before we can tell whether features are homologous, we have to know whether they came from a common ancestor.

Teacher: That's right.

Student (scratching head): I must be missing something. It sounds as though you're saying that we know features are derived from a common ancestor because they're derived from a common ancestor. Isn't that circular reasoning?

At this point, the overburdened teacher might simply end the discussion and move on to something else. But science education would be better served if he or she acknowledged the problem and took some time to analyze it in class. Instead of being told to memorize a circular argument, students might be encouraged to think about the difference between theory and evidence, and how to compare the two.

In the process, they might become not only better scientists, but also better citizens.

CHAPTER 5

Haeckel's Embryos

D arwin knew that the Cambrian fossil record was a serious problem for his theory. He also knew that without a mechanism to explain how homologies were produced, his identification of archetypes with common ancestors remained open to challenge. Thus it seemed to him that neither the fossil record nor homologous structures supported his theory as conclusively as the evidence from embryology.

"It seems to me," Darwin wrote in *The Origin of Species*, "the leading facts in embryology, which are second to none in importance, are explained on the principle of variations in the many descendants from some one ancient progenitor." And those leading facts, according to him, were that "the embryos of the most distinct species belonging to the same class are closely similar, but become, when fully developed, widely dissimilar." Reasoning that "community in embryonic structure reveals community of descent," Darwin concluded that early embryos "show us, more or less completely, the condition of the progenitor of the whole group in its adult state." In other words, similarities in early embryos not only demonstrate that they are descended from a common ancestor, but also reveal what that ancestor looked like.

Darwin considered this "by far the strongest single class of facts in favor of" his theory.

Darwin was not an embryologist, so he relied for his evidence on the work of others. One of those was German biologist Ernst Haeckel (1834-1919). Darwin wrote in *The Origin of Species* that Professor Haeckel "brought his great knowledge and abilities to bear on what he calls phylogeny, or the lines of descent of all organic beings. In drawing up the several series he trusts chiefly to embryological characters."

Haeckel made many drawings, but his most famous were of early vertebrate embryos. Haeckel drew embryos from various classes of vertebrates to show that they are virtually identical in their earliest stages, and become noticeably different only as they develop. (Figure 5-1) It was this pattern of early similarity and later difference that Darwin found so convincing in *The Origin of Species*. Thus "it is probable, from what we know of the embryos of mammals, birds, fishes and reptiles, that these animals are the modified descendants of some ancient progenitor." In *The Descent of Man*, Darwin extended the inference to humans: "The [human] embryo itself at a very early period can hardly be distinguished from that of other members of the vertebrate kingdom." Since humans and other vertebrates "pass through the same early stages of development,... we ought frankly to admit their community of descent."

Haeckel's embryos seem to provide such powerful evidence for Darwin's theory that some version of them can be found in almost every modern textbook dealing with evolution. Yet biologists have known for over a century that Haeckel *faked* his drawings; vertebrate embryos never look as similar as he made them out to be. Furthermore, the stage Haeckel labeled the "first" is actually midway through development; the similarities he exaggerated are preceded by striking differences in earlier stages of development. Although you might never know it from reading biology textbooks, Darwin's "strongest single class of facts" is a classic example of how evidence can be twisted to fit a theory.

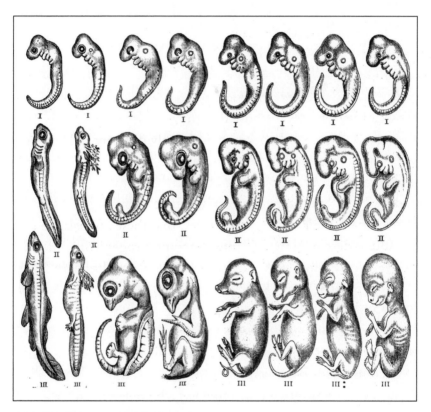

FIGURE 5-1   Haeckel's embryos.

The embryos are (left to right) fish, salamander, tortoise, chick, hog, calf, rabbit, and human. Note that only five of the seven vertebrate classes are represented, and that half the embryos are mammals. This version of Haeckel's drawings is from George Romanes's 1892 book, *Darwinism Illustrated.*

Will the real embryologist please stand up?

Before the publication of *The Origin of Species*, Europe's most famous embryologist was not Ernst Haeckel, but Karl Ernst von Baer (1792-1876). Trained in physics and biology, von Baer had published his major work in embryology by the mid-1830s. That work included four generalizations that became important in subsequent controversies over evolution.

Von Baer's first two generalizations were intended to refute "preformationism," the old idea that embryos are simply minia-ture adults. If preformationism were true, then every embryo would show the distinctive adult characteristics of its species right from the start. But von Baer pointed out that "the more general characters of a large group of animals appear earlier in their embryos than the more special characters."

The second two generalizations were intended to refute the "law of parallelism" which was being promoted by two of von Baer's contemporaries, Johann Friedrich Meckel and Étienne Serres. According to the evolutionary parallelism of Meckel and Serres, the embryos of higher organisms pass through the adult forms of lower organisms in the course of their development. But von Baer noted that "the embryo of a higher form never resembles any other form, but only its embryo."

Although von Baer's generalizations were called "laws," they were actually summaries of empirical observations. They were intended to show that two other "laws"—preformationism and parallelism—did not fit the evidence, and thus should be aban-doned. As a research embryologist, von Baer emphasized the importance of careful observation. It was this that led to his dis-covery of the tiny mammalian egg cell—his principal claim to scientific fame.

Although von Baer accepted the possibility of limited trans-formation of species at lower levels of the biological hierarchy, he saw no evidence for the large-scale transformations proposed by Darwin. For example, von Baer did not believe that the various classes of vertebrates (e.g., fishes, amphibians, reptiles, birds, and mammals) were descended from a common ancestor. According to historian of science Timothy Lenoir, von Baer feared that Darwinists had "already accepted the Darwinian evolutionary hypothesis as true before they set to the task of observing embryos."

So von Baer rejected the evolutionary parallelism of Meckel and Serres, and the large-scale transformations proposed by Darwin. Yet Darwin ended up citing him as the source of the "strongest single class of facts" supporting his theory of evolution.

Darwin's misuse of von Baer

Darwin apparently never read von Baer, who wrote in German. The first two editions of *The Origin of Species* cited a passage of von Baer's that had been translated by Thomas Henry Huxley, but Darwin mistakenly attributed the passage to Louis Agassiz. Only in the third and subsequent editions did he mention von Baer.

Darwin wrote: "Generally the embryos of the most distinct species belonging to the same class are closely similar, but become, when fully developed, widely dissimilar. A better proof of this latter fact cannot be given than the statement by von Baer that 'the embryos of mammals, birds, lizards and snakes, and probably [turtles] are in their earliest states exceedingly like one another.... In my possession are two little embryos in spirit, whose names I have omitted to attach, and at present I am quite unable to say to what class they belong. They may be lizards or small birds, or very young mammals, so complete is the similarity in the mode of formation of the head and trunk in these animals.'"

When von Baer wrote this he may have been exaggerating, because in fact the embryos of lizards, birds, and mammals *can* be distinguished at an early age. And the embryos of other vertebrate classes, such as fishes and amphibians, look even more different. In any case, von Baer knew that embryos never look like the adult of another species, and he saw no evidence for Darwin's theory that the various classes of vertebrates shared a common ancestor. Yet several pages after citing von Baer as his authority in these matters, Darwin claimed that "it is probable, from what we know of the embryos of mammals, birds, fishes and reptiles, that these animals are the modified descendants of some ancient progenitor," and that "with many animals the embryonic or larval stages show us, more or less completely, the condition of the progenitor of the whole group in its adult state."

This last claim is exactly what von Baer's second two laws denied. In other words, Darwin cited von Baer as the source of his embryological evidence, but at the crucial point Darwin distorted that evidence to make it fit his theory. Von Baer lived long enough to object to Darwin's misuse of his observations, and he was a strong critic of Darwinian evolution until his death in 1876. But Darwin persisted in citing him anyway, making him look like a supporter of the very doctrine of evolutionary parallelism he explicitly rejected.

In what historian of science Frederick Churchill calls "one of the ironies of nineteenth-century biology," von Baer's view "was confounded with and then transformed into an evolutionary form of the law of parallelism." Naturalist Fritz Müller (whom Darwin also cited) "encouraged the confusion," but it was Müller's student, Ernst Haeckel, who "dramatized the obfuscation" and became its most ardent promoter.

Haeckel's biogenetic law

Haeckel coined the terms "ontogeny" to designate the embryonic development of the individual, and "phylogeny" to designate the evolutionary history of the species. He maintained that embryos "recapitulate" their evolutionary history by passing through the adult forms of their ancestors as they develop. When new features evolve they are tacked on to the end of development, in a process Stephen Jay Gould calls "terminal addition," making ancestral forms appear earlier in development than more recently evolved features. Haeckel called this the "biogenetic law" and summarized it in the now-famous phrase, "ontogeny recapitulates phylogeny."

Von Baer's laws and Haeckel's biogenetic law are very different. The former were based on empirical observations and intended to refute theories that didn't fit the evidence, while the latter was deduced from evolutionary theory rather than inferred from evidence. "The recapitulation theory," wrote British zoologist Adam Sedgwick in 1909, "originated as a deduction from the evolution theory and as a deduction it still remains." Ten years later, American embryologist Frank Lillie likewise acknowledged that recapitulation is a logical consequence of evolution rather than an empirical inference, though he was inclined to accept it anyway. Lillie reasoned that since "the basis of any theory of descent is heredity, and it must be recognized that ontogenies are inherited, the resemblance between the individual history and the phylogenetic history necessarily follows."

So from the very beginning, Haeckel's biogenetic law was a theoretical deduction rather than an empirical inference. It exerted considerable influence in the late nineteenth and early twentieth centuries, but by the 1920s it was losing favor. According to Stephen Jay Gould, "the biogenetic law fell only when it became unfashionable." Historian of science Nicholas Rasmussen agrees. Certainly, it did not fall because new discoveries contradicted it. As Rasmussen puts it: "All the important evidence called upon in the rejection of the biogenetic law was there from the first days of the law's acceptance."

Resurrecting recapitulation

Nevertheless, some twentieth-century American and British embryologists attempted to salvage what they considered an element of truth in Haeckel's law. Lillie knew that Haeckel's law was empirically false. He also knew that von Baer's laws had only limited applicability, because "it never happens that the embryo of any definite species resembles in its entirety the adult of a lower species, nor even the embryo of a lower species; its organization is specific at all stages from the [egg] on, so that it is possible without any difficulty to recognize the order of animals to which a given embryo belongs." Nevertheless, on theoretical grounds Lillie affirmed some sort of parallelism between ontogeny and phylogeny.

In 1922 British embryologist Walter Garstang criticized Haeckel's biogenetic law as "demonstrably unsound," because "ontogenetic stages afford not the slightest evidence of the specially adult features of the ancestry." According to Garstang,

Haeckel's theory that newly evolved features are simply tacked onto the end of development makes no sense: "A house is not a cottage with an extra story on the top. A house represents a higher grade in the evolution of a residence, but the whole building is altered—foundations, timbers, and roof—even if the bricks are the same." Nevertheless, Garstang (like Lillie) maintained on theoretical grounds that there must be a general correspondence between ontogeny and phylogeny, and that in this "original and general sense"—which Garstang attributed to Meckel—"recapitulation is a fact." So Garstang and Lillie knew that the biogenetic law did not fit the evidence, but because of their belief in Darwinian evolution they were convinced that some form of recapitulationism had to be true.

From 1940 to 1958 British embryologist Gavin de Beer published three editions of a book on embryology and evolution in which he criticized Haeckel's biogenetic law. "Recapitulation," wrote de Beer, "i.e., the pressing back of adult ancestral stages into early stages of development of descendants, does not take place." But the problem was not merely the claim that *adult* forms are recapitulated, since "variations of evolutionary significance can and do arise at the earliest stages of development." In other words, the earliest stages of development show important differences, contrary to Darwin's belief that they are the most similar. De Beer concluded that recapitulation is "a mental straitjacket" that "has thwarted and delayed" embryological research.

Yet if organisms are descended from a common ancestor, it seems reasonable to expect ontogeny to provide evidence of phylogeny. Recapitulation in some sense is a logical consequence of Darwinian evolution. The question is: What sense? In discussions of development and evolution, two views keep recurring. Both are found in Darwin's *Origin of Species*:

I. The earliest stages of embryos are more similar than their later stages. In Darwin's words: "The embryos of the most distinct species belonging to the same class are closely similar, but become, when fully developed, widely dissimilar."

II. Embryos pass through the adult forms of their ancestors as they develop. In Darwin's words: "With many animals the embryonic or larval stages show us, more or less completely, the condition of the progenitor of the whole group in its adult state."

The first view is von Baer's, though he would not have extended it beyond the level of classes. Modern Darwinists sometimes call it "von Baerian recapitulation," though this is actually an oxymoron—on a par with "Copernican geocentrism" or "Darwinian creationism." The second view is Haeckel's biogenetic law, and is thus called "Haeckelian recapitulation."

Both views are empirically false. Yet throughout the twentieth century they have periodically risen, phoenix-like, from the ashes of empirical disconfirmation. Since both are frequently enlisted in support of Darwinian evolution, it is often difficult to tell them apart. And as we shall see below, in one of the most bizarre twists of all, both are now illustrated with the same set of faked drawings.

Haeckel's embryo drawings

Haeckel produced many drawings of vertebrate embryos to illustrate his biogenetic law. The drawings show vertebrate embryos that look very much alike at their earliest stage. (Figure 5-1, top row) In fact, the embryos look *too* much alike. According to historian Jane Oppenheimer, Haeckel's "hand as an artist altered what he saw with what should have been the eye of a more accurate beholder. He was more than once, often justifiably, accused of scientific falsification, by Wilhelm His and many others."

In some cases, Haeckel used the same woodcut to print embryos that were supposedly from different classes. In others, he doctored his drawings to make the embryos appear more alike than they really were. Haeckel's contemporaries repeatedly criticized him for these misrepresentations, and charges of fraud abounded in his lifetime.

Whether or not Haeckel was guilty of fraud—that is, deliberate deception—there is no doubt that his drawings misrepresent vertebrate embryos. First, he chose only those embryos that came closest to fitting his theory. Although there are seven classes of vertebrates (jawless fishes, cartilaginous fishes, bony fishes, amphibians, reptiles, birds, and mammals), Haeckel showed only five, omitting jawless and cartilaginous fishes entirely. Furthermore, to represent amphibians he used a salamander rather than a frog, which looks very different. Finally, half of his embryos are mammals, and all of these are from one order (placentals); other mammalian orders (egg-laying monotremes and pouch-brooding marsupials) are omitted. Thus, Haeckel began with a biased sample.

Even the embryos he chose are distorted to fit his theory. British embryologist Michael Richardson noted in 1995 that the top row of embryos in Haeckel's drawings is "not consistent with other data on the development of these species." Richardson concluded: "These famous images are inaccurate and give a misleading view of embryonic development." In 1997 Richardson and an international team of experts compared Haeckel's embryos with photographs of actual embryos from all seven classes of vertebrates, showing quite clearly that Haeckel's drawings misrepresent the truth.

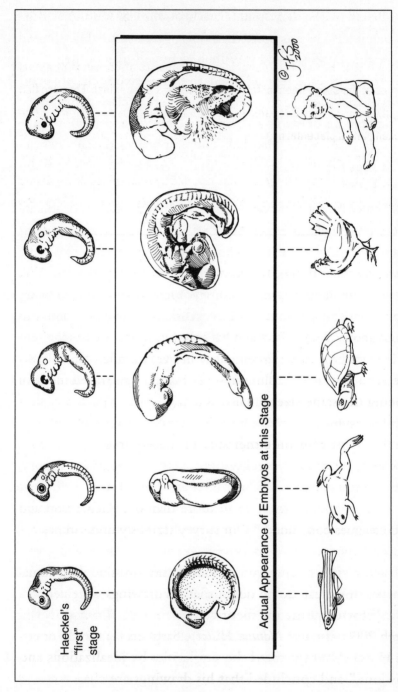

Haeckel's "first" stage

Actual Appearance of Embryos at this Stage

FIGURE 5-2 A comparison of haeckel's drawings with actual vertebrate embryos.

FIGURE 5-2 **A comparison of haeckel's drawings with actual verte-brate embryos.**

The top row is Haeckel's. The middle row consists of drawings of actual embryos at the stage Haeckel falsely claimed was the earliest. They are (left to right): a bony fish (zebrafish); an amphibian (frog); a reptile (turtle); a bird (chicken); and a placental mammal (human). To represent amphibians Haeckel used a salamander, which fits his theory better than a frog; a frog is used here to highlight this fact. Other groups not included by Haeckel (such as jawless and cartilaginous fishes, and monotreme and marsupial mammals) are signifi-cantly different from the embryos shown here.

Among other things, Richardson and his colleagues found that "there is great variation in embryonic morphology" among amphibians, but Haeckel chose a salamander that happened to fit his theory. Richardson and his colleagues also found that ver-tebrate embryos vary tremendously in size, from less than 1 mil-limeter to almost 10 millimeters, yet Haeckel portrayed them all as being the same size. Finally, Richardson and his colleagues found considerable variation in the number of somites—repeti-tive blocks of cells on either side of the embryo's developing backbone. Although Haeckel's drawings (Figure 5-1, top row) show approximately the same number of somites in each class, actual embryos vary from 11 to more than 60. Richardson and his colleagues concluded: "Our survey seriously undermines the credibility of Haeckel's drawings."

When Haeckel's embryos are viewed side-by-side with actual embryos, there can be no doubt that his drawings were deliber-ately distorted to fit his theory. (Figure 5-2) Writing in the March 2000, issue of *Natural History*, Stephen Jay Gould noted that Haeckel "exaggerated the similarities by idealizations and omissions," and concluded that his drawings are characterized

by "inaccuracies and outright falsification." Richardson, inter-
viewed by *Science* after he and his colleagues published their
now-famous comparisons between Haeckel's drawings and actual
embryos, put it bluntly: "It looks like it's turning out to be one
of the most famous fakes in biology."

So Haeckel's drawings are fakes, and they misrepresent the
embryos they purport to show. But they are fakes in another
sense, too. Darwin based his inference of common ancestry on
the belief that the *earliest* stages of embryo development are the
most similar. Haeckel's drawings, however, omit the earliest
stages entirely, and start at a point midway through development.
The earlier stages are much different.

The earliest stages in vertebrate embryos are not the most similar

When an animal egg is fertilized, it first undergoes a process
called "cleavage," during which it subdivides into hundreds or
thousands of separate cells without growing in overall size. At the
end of cleavage, the cells begin to move and rearrange themselves
in a process known as "gastrulation." Gastrulation, even more
than cleavage, is responsible for establishing the animal's general
body plan (e.g., insect or vertebrate) and for generating basic
tissue types and organ systems (e.g., skin, muscles, and gut).
British embryologist Lewis Wolpert has written that "it is not
birth, marriage, or death, but gastrulation which is truly 'the
important event in your life'."

Yet only after cleavage and gastrulation does a vertebrate
embryo reach the stage which Haeckel labeled the "first." If it
were true (as Darwin and Haeckel claimed) that vertebrates are
most similar in the earliest stages of their development, then the
various classes would be most similar during cleavage and
gastrulation. Yet a survey of five classes (bony fish, amphibian, reptile, bird and mammal) reveals that this is not the case. (Figure 5-3)

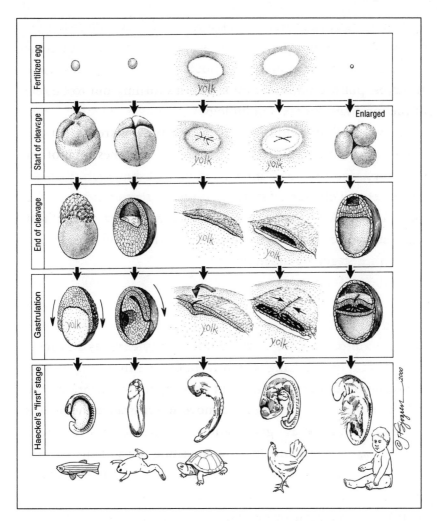

FIGURE 5-3 **Early stages in vertebrate embryos.**

Drawings of early embryonic stages in five classes of vertebrates. The stages are (top to bottom): fertilized egg; early cleavage; end of cleavage; gastrulation; and Haeckel's "first" stage. The fertilized eggs are drawn to scale relative to each other, while the scales of the succeeding stages are normalized to facilitate comparisons. The embryos are (left to right): bony fish (zebrafish), amphibian (frog), reptile (turtle), bird (chicken), and mammal (human).

Differences among the five classes are evident even in the fertilized eggs. Zebrafish and frog eggs are about a millimeter in diameter; turtles and chicks start out as discs 3 or 4 millimeters in diameter that rest on top of a large yolk; while the human egg is only about a tenth of a millimeter in diameter. (Figure 5-3, top row) The earliest cell divisions in zebrafish, turtle, and chick embryos are somewhat similar, but in most frogs they penetrate the yolk. Mammals are completely different, however, since one of the second cleavage planes is at a right angle to the other. (Figure 5-3, second row) Continued cleavage in the other four classes produces a stable arrangement of cells, but mammalian embryos become a jumbled mass.

At the end of cleavage, the cells of the zebrafish embryo form a large cap on top of the yolk; in the frog they form a ball with a cavity; in the turtle and chick they form a thin, two-layered disc on top of the yolk; and in humans they form a disc within a ball. (Figure 5-3, third row) Cell movements during gastrulation are very different in the five classes: In zebrafish the cells crawl down the outside of the yolk; in frogs they move as a coherent sheet through a pore into the inner cavity; and in turtles, chicks, and humans they stream through a furrow into the hollow interior of the embryonic disc. (Figure 5-3, fourth row)

If the implications of Darwin's theory for early vertebrate development were true, we would expect these five classes to be most similar as fertilized eggs; slight differences would appear during cleavage, and the classes would diverge even more during gastrulation. What we actually observe, however, is that the eggs of the five classes start out noticeably different from each other; the cleavage patterns in four of the five classes show some general similarities, but the pattern in mammals is radically different. In the gastrulation stage, a fish is very different from an amphibian, and both are very different from reptiles, birds, and mammals, which are somewhat similar to each other. Whatever pattern can be discerned here, it is certainly *not* a pattern in which the earliest stages are the most similar and later stages are more different.

The dissimilarity of early embryos is well-known

The dissimilarity of early vertebrate embryos has been known to biologists for over a century. Embryologist Adam Sedgwick pointed out in 1894 that von Baer's law of early similarity and later difference is "not in accordance with the facts of development." Comparing a dog-fish with a fowl (i.e., a chicken), Sedgwick wrote: "There is no stage of development in which the unaided eye would fail to distinguish between them with ease." Even more to the point: "If von Baer's law has any meaning at all, surely it must imply that animals so closely allied as the fowl and duck would be indistinguishable in the early stages of development;... yet I can distinguish a fowl and a duck embryo on the second day." It is "not necessary to emphasize further these embryonic differences," Sedgwick continued, because "every embryologist knows that they exist and could bring forward innumerable instances of them. I need only say with regard to them that *a species is distinct and distinguishable from its allies from the very earliest stages all through the development.*" (emphasis in the original)

Modern embryologists confirm this. William Ballard wrote in 1976 that it is "only by semantic tricks and subjective selection of evidence," by "bending the facts of nature," that one can argue that the cleavage and gastrulation stages of vertebrates "are more alike than their adults." The following year Erich Blechschmidt noted: "The early stages of human embryonic development are distinct from the early development of other species." And in 1987 Richard Elinson reported that frogs, chicks, and mice "are radically different in such fundamental properties as egg size, fertilization mechanisms, cleavage patterns, and [gastrulation] movements."

Surprisingly, after developing quite differently in their early stages, vertebrate embryos become somewhat similar midway through development. It is this midway point that Haeckel chose as the "first" stage for his drawings. Although he greatly exaggerated the similarities at this stage, some similarities are there. Classical embryologists called this midpoint the "tailbud stage." In 1976 William Ballard called it the "pharyngula" because of the paired ridges and pouches on either side of the pharynx. Klaus Sander proposed in 1983 to call it the "phylotypic stage," since it is here that the various classes first exhibit the characteristics common to all vertebrates.

Some developmental biologists, however, point out that the midpoint at which vertebrate embryos are most similar is spread out over several stages. "The phylotypic point is neither a point nor a stage," wrote Denis Duboule in 1994, "but rather, a succession of stages." And according to Michael Richardson "the phylotypic stage is a misleading concept that needs to be reassessed," because "in vertebrates, body plan characters develop over a long range of different stages, not just at one stage."

Nevertheless, no one doubts that vertebrate embryos start out looking very different, converge in appearance midway through development (though not at the same time), then become increasingly more different as they continue toward adulthood. Duboule uses the metaphor of a "developmental egg-timer" to describe this pattern, while Rudolf Raff calls it the "develop-mental hourglass." (Figure 5-4) Although von Baer's laws do not apply to embryonic stages before the middle of the develop-mental hourglass, they do appear to be roughly applicable to later stages. As Raff wrote in 1996: "It should be noted that von Baer's laws provide an incomplete description of develop-ment.... In fact, he was dealing only with the later half of ontogeny."

A paradox for Darwinian evolution

But if von Baer's laws apply only to the second half of ontogeny, descent with modification is deprived of what Darwin believed to be "the strongest single class of facts" in favor of it. Accord-ing to Darwin, it was the similarity of embryos *in their earliest stages* that provided evidence for common descent. The actual pattern—early differences followed by similarities, then differ-ences again—is quite unexpected in the context of Darwinian evolution. Instead of providing support for Darwin's theory, the embryological evidence presents it with a paradox.

Recently, some embryologists have sought to explain the paradox by proposing that early development evolves much more easily than anyone expected. According to Gregory Wray, differences in early development indicate that "profound changes in developmental mechanisms can evolve quite rapidly." Rudolf Raff suggests that "the evolutionary freedom of early ontogenetic stages is significant in providing novel develop-mental patterns and life histories." Whatever the merit of such proposals may be, it is clear that they start by *assuming* Darwin-ian evolution, then read that back into the embryological evidence.

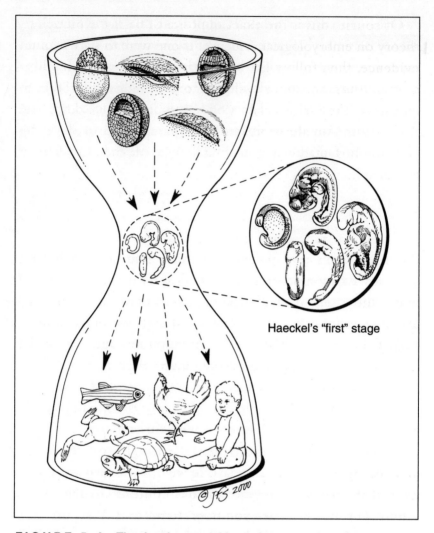

FIGURE 5-4 The developmental hourglass.

The vertical axis represents developmental time, from top to bottom; the horizontal axis represents morphological diversity. Vertebrate embryos start out looking very different, then superficially converge midway through development at the "pharyngula" or "phylotypic" stage, before diverging into their adult forms.

Of course, this is the exact opposite of basing evolutionary theory on embryological evidence. If one were to start with the evidence, then follow Darwin's reasoning about the implications of development for evolution, one would presumably conclude that the various classes of vertebrates are *not* descended from a common ancestor, but had separate origins. Since this conclusion is unacceptable to people who have already decided that Darwin's theory is true, they cannot take the embryological evidence at face value, but must re-interpret it to fit the theory.

So we have come back to our starting-point. Von Baer objected to nineteenth-century Darwinists because they accepted evolutionary theory before they even began looking at embryos. Many modern Darwinists haven't changed. It doesn't matter how much the embryological evidence conflicts with evolutionary theory—the theory, it seems, must not be questioned. This is why, despite repeated disconfirmation, Haeckel's biogenetic law and faked drawings haven't gone away.

Haeckel is dead. Long live Haeckel.

Since Darwin's theory is affirmed regardless of the evidence, and "ontogeny recapitulates phylogeny" is a logical deduction from that theory, biology textbooks continue to teach it— though they usually attach von Baer's name to it. Thus the 1975 edition of B. I. Balinsky's classic textbook, *Introduction to Embryology,* includes this amazing passage: "Von Baer's law... can be reinterpreted in the light of evolutionary theory. In its new form the law is known as the biogenetic law of Müller-Haeckel." According to von Baer's law, the book continues, "features of ancient origin develop early in ontogeny; features of newer origin develop late. Hence, the ontogenetic development presents the various features of the animal's organization in the same sequence as they evolved during the phylogenetic development. *Ontogeny is a recapitulation of phylogeny.*" (emphasis in the original)

It is difficult to imagine how any history of the biogenetic law could be more distorted than this. Yet the distortion is perpetuated in many modern biology textbooks. And as if this weren't bad enough, some textbooks even use Haeckel's faked drawings to illustrate von Baer's law.

For example, Haeckel's drawings are reproduced in the 1998 edition of Douglas Futuyma's advanced college textbook, *Evolutionary Biology*, but the figure caption doesn't mention Haeckel; instead, it describes the drawings as "an illustration of von Baer's law." And the most recent edition of *Invitation to Biology*, by Helena Curtis and Sue Barnes, reproduces the top two lines of Haeckel's drawings with the following caption: "These drawings are based on the work of the nineteenth-century embryologist Karl Ernst von Baer."

Yet falsely attributing Haeckel's ideas and drawings to von Baer is not the most serious offense in these textbooks. That distinction goes to their use of Haeckel's drawings to misrepresent the embryological evidence. As we have seen, Haeckel's drawings are misleading in three ways: (1) they include only those classes and orders that come closest to fitting Haeckel's theory; (2) they distort the embryos they purport to show; and (3) most seriously, they entirely omit earlier stages in which vertebrate embryos look very different.

Haeckel's drawings appear not only in Futuyma's book and the book by Curtis and Barnes, but also in the latest edition of *Molecular Biology of the Cell*, by National Academy of Sciences President Bruce Alberts and his colleagues. "Early developmental stages of animals whose adult forms appear radically different are often surprisingly similar," the Alberts textbook claims, and neo-Darwinian mechanisms explain why "embryos of different species so often resemble each other in their early stages and, as they develop, seem sometimes to replay the steps of evolution."

Many textbooks use slightly redrawn versions of Haeckel's embryos. One example is the 1999 edition of Peter Raven and George Johnson's *Biology*, which accompanies its drawings with the following caption: "Notice that the early embryonic stages of these vertebrates bear a striking resemblance to each other." The text also informs students: "Some of the strongest anatomical evidence supporting evolution comes from comparisons of how organisms develop. In many cases, the evolutionary history of an organism can be seen to unfold during its development, with the embryo exhibiting characteristics of the embryos of its ancestors."

Other examples include the 1998 edition of Cecie Starr and Ralph Taggart's *Biology: The Unity and Diversity of Life*, which accompanies its drawings with the mis-statement that "the early embryos of vertebrates strongly resemble one another;" the latest edition of James Gould and William Keeton's *Biological Science*, which reports: "One fact of embryology that pushed Darwin toward the idea of evolution is that the early embryos of most vertebrates closely resemble one another;" and Burton Guttman's 1999 textbook, *Biology*, which accompanies its redrawn version of Haeckel's embryos with the following: "An animal's embryonic development holds clues to the forms of its ancestors."

Some textbooks, instead of reproducing or redrawing Haeckel's embryos, use actual *photos*. Sylvia Mader's 1998 *Biology*, for example, includes photos of chick and pig embryos, accompanied by the caption: "At these comparable early developmental stages, the two have many features in common, although eventually they are completely different animals. This is evidence that they evolved from a common ancestor." Mader's use of actual photos instead of faked drawings is a step in the right direction, but the embryological evidence is still being misrepresented. As we have seen, Haeckel's distortions of embryos in mid-development was just *one* of his misrepresentations; the others were his biased selection of classes and orders that fit his theory, and his omission of earlier stages. Both of these misrepresentations are perpetuated— recapitulated, one might say—by Mader.

The 1999 edition of Campbell, Reece, and Mitchell's *Biology* also uses photos of actual embryos that mislead students. Like Mader's book, this one compares a mammal with a chick, which just happens to look more like a mammal than any other class of vertebrate at that stage. Although the textbook warns students that "the theory of recapitulation is an overstatement," it also tells them that "ontogeny can provide clues to phylogeny."

Is a human embryo like a fish?

The use of embryo photos to mislead people about recapitulation is not limited to textbooks. The November 1996 issue of *Life* magazine contains spectacular photos of an embryonic human, macaque monkey, lemur, pig and chick. The pictures were the work of photographer Lennart Nilsson, and the accompanying text was written by Kenneth Miller.

Miller describes the development of the human embryo as a "microscopic trip through evolutionary time," though he rejects Haeckel's biogenetic law that a human "on its way to birth becomes a fish, an amphibian and so on up the evolutionary ladder." Recapitulationism, according to Miller, "provides an example of how appearances can deceive even eminent scientists."

Yet Miller also describes how human embryos "grow fin-like appendages and something very much like gills." These "gill-like" features are "the legacy of a primitive fish," and this "is some of the most compelling evidence of evolution since Charles Darwin published *The Origin of Species* in 1859."

Miller is not the only one who claims to see "gill-like" features in human embryos. According to Curtis and Barnes's *Invitation to Biology*, "early [vertebrate] embryos are almost indistinguishable. All have prominent gill pouches." Gould and Keeton's *Biological Science* informs students that "telltale traces of their genealogy are obvious in vertebrates... Human embryos, for instance, have gill pouches." Raven and Johnson's *Biology* claims that "early in their development, human embryos possess gill slits, like a fish." And Futuyma's *Evolutionary Biology* likewise states: "Early in development, human embryos are almost indistinguishable from those of fishes, and briefly display gill slits."

All of these statements, however, are versions of Haeckel's biogenetic law. All of them project evolutionary theory back into the embryological evidence, and distort that evidence to make it fit the theory. The true picture looks quite different.

"Gill slits" are not gill slits

Midway through development, all vertebrate embryos possess a series of folds in the neck region, or pharynx. The convex parts of the folds are called pharyngeal "arches" or "ridges," and the concave parts are called pharyngeal "clefts" or "pouches." But pharyngeal folds are not gills. They're not even gills in pharyngula-stage *fish* embryos.

In a fish, pharyngeal folds later develop into gills, but in a reptile, mammal, or bird they develop into other structures entirely (such as the inner ear and parathyroid gland). In reptiles, mammals, and birds, pharyngeal folds are never even rudimentary gills; they are *never* "gill-like" except in the superficial sense that they form a series of parallel lines in the neck region. According to British embryologist Lewis Wolpert: "A higher animal, like the mammal, passes through an embryonic stage when there are structures that resemble the gill clefts of fish. But this resemblance is illusory and the structures in mammalian embryos only resemble the structures in the *embryonic* fish that will give rise to gills."

In other words, there is no *embryological* reason to call pharyngeal pouches "gill-like." The only justification for that term is the theoretical claim that mammals evolved from fish-like ancestors. Swiss embryologist Günter Rager explains: "The concept 'pharyngeal arches' is purely descriptive and ideologically neutral. It describes folds which appear [in the neck] region.... In man, however, gills do never exist."

The only way to see "gill-like" structures in human embryos is to read evolution into development. But once this is done, development cannot be used as evidence for evolution without plunging into circular reasoning—like that used to infer common ancestry from the neo-Darwinian concept of homology. (Chapter 4) To put it bluntly: There is no way "gill-slits" in human embryos can logically serve as evidence for evolution.

Despite protestations that nobody any longer believes in Haeckelian recapitulation, here it is again. Gills are not embryonic structures, not even in fish. "Seeing" them in other classes of vertebrates is to read an adult structure back into the embryo.

So recapitulation continues to rear its ugly head. Although biologists have known for over a century that it doesn't fit the evidence, and although it was supposedly discarded in the 1920s, recapitulation continues to distort our perceptions of embryos.

Furthermore, although biologists have also known for over a century that Haeckel's drawings are fakes, and that the earliest stages in vertebrate development are not the most similar, textbooks continue to use those drawings (or almost equally misleading photos) to convince unsuspecting students that Darwin's theory rests on embryological evidence.

Since 1997 when Richardson and his colleagues reminded biologists that Haeckel's embryos misrepresent the truth, Darwinists have come under increasing criticism for continuing to use them. Just recently, Douglas Futuyma and Stephen Jay Gould have been moved to respond to those criticisms.

Atrocious!

In February 2000 textbook-writer Douglas Futuyma posted a message to a Kansas City internet forum in response to a critic who had accused him of lying by using Haeckel's embryos in his 1998 textbook, *Evolutionary Biology*. In his defense, Futuyma explained that before reading the critic's accusation he had been unaware of the discrepancies between Haeckel's drawings and actual vertebrate embryos. Only after consulting a developmental biologist had he learned about the recent work of Richardson and his colleagues.

So Futuyma, a professional evolutionary biologist and author of a graduate-level textbook, did not know about Haeckel's faked drawings—a confession of ignorance not likely to inspire much confidence in the quality of our biology textbooks. But now he knows that "Haeckel was inaccurate and misleading," and he said he would take this into account in future editions of his book.

Futuyma maintained, however, that even though Haeckel had exaggerated their similarities "the various embryos really are very similar—we are talking about pretty minor differences." He argued that "Haeckel's inaccuracies, whether intended to deceive or not, are trivial compared to the overall message." The message, according to Futuyma, is that what he calls von Baer's law is true: "Bird and mammal embryos are really more similar than the adults." For example, "all the vertebrate embryos... *really do* have gill slits." (emphasis in the original)

In the March 2000 issue of *Natural History* magazine, Stephen Jay Gould responded to Michael Behe, a biologist who had criticized Haeckel's embryos in the August 13, 1999, *New York Times*. Gould acknowledged that Haeckel faked his drawings. "To cut to the quick of this drama," Gould wrote, "Haeckel had exaggerated the similarities by idealizations and omissions. He also, in some cases—in a procedure that can only be called fraudulent—simply copied the same figure over and over again."

Unlike Futuyma, however, Gould admitted that he already knew this; in fact, he had known about it for more than twenty years. (As a historian of science, Gould wrote a major book on the subject in 1977, *Ontogeny and Phylogeny*.) He blamed recent news reports for sensationalizing the story by giving the impression "that Richardson had discovered Haeckel's misdeed for the first time." Gould continued: "Tales of scientific fraud excite the imagination for good reason. Getting away with this academic equivalent of murder and then being outed a century after your misdeeds makes even better copy."

But if biologists have known all along that Haeckel's drawings were faked, then why are they still used? Gould laid the blame at the feet of textbook-writers, blasting them for "dumbing down" their subject matter to the point of making it inaccurate. "We do, I think, have the right," he wrote, "to be both astonished and ashamed by the century of mindless recycling that has led to the persistence of these drawings in a large number, if not a majority, of modern textbooks."

So Gould blames the textbook writer, while the textbook writer pleads ignorance. Both of them, however, are quick to criticize "creationists." "Note that science is a self-correcting process," wrote Futuyma in response to his Kansas critic, "unlike creationist critiques of science; evolutionary biologists themselves reveal inaccuracies in the earlier literature of their field." And Gould blames creationists for capitalizing on the work of Richardson and his colleagues by making the "ersatz" and "sensationalist" charge that "a primary pillar of Darwinism, and of evolution in general, had been revealed as fraudulent after more than a century" of uncritical acceptance.

But it was Futuyma who mindlessly recycled Haeckel's embryos in several editions of his textbook, until a "creationist" criticized him for it. And it was Gould who (despite having known the truth for over twenty years) kept his mouth shut until a "creationist" (actually, a fellow biologist) exposed the problem. And all that time, Gould was letting his colleagues become accessories to what he himself calls "the academic equivalent of murder."

CHAPTER 6

Archaeopteryx:
The Missing Link

When Charles Darwin published *The Origin of Species* in 1859, he acknowledged that the fossil record was a serious problem for his theory. "By the theory of natural selection," he wrote, "all living species have been connected with the parent-species of each genus, by differences not greater than we see between the natural and domestic varieties of the same species at the present day." As a consequence, "the number of intermediate and transitional links, between all living and extinct species, must have been inconceivably great." Yet in 1859 those transitional links had not been found.

Darwin attributed their absence to "the imperfection of the geological record." He argued that most organisms were never preserved, or if preserved were subsequently destroyed, so that "we have no right to expect to find, in our geological formations, an infinite number of those transitional forms which, on our theory, have connected all the past and present species of the same group into one long and branching chain of life. We ought only to look for a few links."

Two years later, in the midst of heated controversy over Darwin's theory, came the dramatic announcement that one of those links had just been found. In 1861 Hermann von Meyer described a fossil that appeared to be intermediate between reptiles and birds. Discovered in a limestone quarry in Solnhofen, Germany, the fossil had wings and feathers; but it also had teeth (unlike any modern bird), a long lizard-like tail, and claws on its wings. Meyer named the newly discovered animal *Archaeopteryx* (meaning "ancient wing").

In 1877 an even more complete specimen of *Archaeopteryx* was discovered. The first specimen ended up in the Natural History Museum in London (and is now known as the "London specimen"), while the second ended up in the Humboldt Museum in Berlin (the "Berlin specimen"). (Figure 6-1) Six other specimens have been found, making a total of eight (though one is just a feather, and one has been lost). But the Berlin *Archaeopteryx* is the most complete and best-preserved, and it has become familiar to millions of people as the missing link that confirmed Darwin's theory.

Yet the role of *Archaeopteryx* as a link between reptiles and birds is very much in dispute. Paleontologists now agree that *Archaeopteryx* is not the ancestor of modern birds, and its own ancestors are the subject of one of the most heated controversies in modern science. The missing link, it seems, is still missing.

The "First Bird"

The Solnhofen limestone, in which all eight specimens of *Archaeopteryx* were discovered, is from the geological period known as the Upper (or Late) Jurassic, about 150 million years ago. This makes *Archaeopteryx* the earliest known bird—or at least, the earliest undisputed bird. Several specimens of it—especially the Berlin specimen—are also among the most beautiful fossils ever found. The Solnhofen limestone is so fine-grained that it is quarried for use in the printing process known as lithography, and it preserved *Archaeopteryx* in exquisite detail—right down to the structure of its feathers.

"To museum curators," write paleontologists Lowell Dingus and Timothy Rowe, "the name *Archaeopteryx* rings like that of Rembrandt, Stradivarius, or Michelangelo." In the words of ornithologist Alan Feduccia, the Berlin *Archaeopteryx* "may well be the most important natural history specimen in existence.... Beyond doubt, it is the most widely known and illustrated fossil

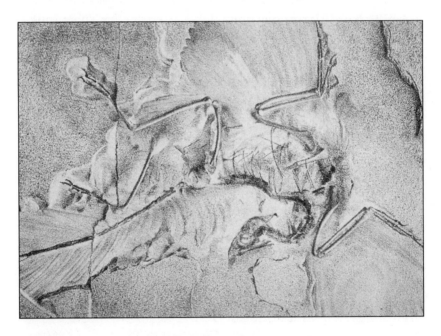

FIGURE 6-1 **The Berlin *Archaeopteryx.***

This is the most complete and well-preserved of the eight known specimens of *Archaeopteryx*. It is owned by the Humboldt Museum in Berlin. (Photo courtesy of the Linda Hall Library, Kansas City, Missouri.)

animal." And to paleontologist Pat Shipman it is "more than the world's most beautiful fossil.... [it is] an icon—a holy relic of the past that has become a powerful symbol of the evolutionary process itself. It is the First Bird."

The iconic status of the First Bird has not gone unchallenged. In 1983 Texas paleontologist Sankar Chatterjee found a fossil from the Late Triassic, about 225 million years ago, which he declared to be "the oldest known fossil bird." When Chatterjee's colleagues examined the fossil, however, they found "roadkill" that was "smushed and smashed and broken." No feathers were present. Some experts even questioned whether all the bones were part of the same animal. Chatterjee has since found other specimens, though none of them have feathers, either. Other paleontologists remain skeptical.

Another kind of challenge to *Archaeopteryx* came in 1986 from British cosmologists Fred Hoyle and Chandra Wickramasinghe. Hoyle and Wickramasinghe claimed that the London specimen had been faked by pressing modern feathers into cement that had been painted onto the fossil of a small dinosaur. British paleontologist Alan Charig and his colleagues showed, however, that the forgery charge was unfounded. Although the significance of *Archaeopteryx* for bird evolution remains controversial, all parties to the current controversy agree that the fossils are genuine.

The missing link

When the first skeleton of *Archaeopteryx* was discovered in 1861, it was widely heralded as the missing link predicted by Darwin's theory. Scientists at the time called it "unimpeachable" evidence for evolution. The enormous gap between reptiles and birds that had previously seemed unbridgeable now seemed to be bridged by a reptile-like bird.

The most striking thing about *Archaeopteryx* is its wonderfully preserved feathers, which are structurally similar to the feathers of modern flying birds. But the animal had toothed jaws like a reptile, rather than a bird-like beak, and it had a long, bony reptile-like tail. It also had claws on its wings, a feature that appears transiently during development in only a few modern birds.

Darwin's ardent defender, Thomas Henry Huxley, helped to publicize *Archaeopteryx*, though he actually regarded another Solnhofen fossil as a more important "missing link" between reptiles and birds. The other fossil was *Compsognathus,* a small, bird-like dinosaur that looked a bit like *Archaeopteryx* but had no feathers. One specimen of *Archaeopteryx* (collected in 1951) in which feathers were not immediately recognized was even misidentified as *Compsognathus* for several years.

Although Huxley regarded *Archaeopteryx* as important evidence for Darwin's theory, he considered *Compsognathus* "a still nearer approximation to the 'missing link' between reptiles and birds," and even suggested that birds had evolved from dinosaurs. He acknowledged, however, that "we have no knowledge of the animals which linked reptiles and birds together historically and genetically," and that fossils "only help us to form a reasonable conception of what those intermediate forms may have been."

In the last edition of *The Origin of Species*, Darwin took note of the recent fossil discoveries that had persuaded many people of the truth of his theory. "Even the wide interval between birds and reptiles," he wrote, "has been shown by [Huxley] to be partially bridged over in the most unexpected manner" by *Archaeopteryx* and *Compsognathus*. Since the latter was the contemporary of the former, however, it couldn't be its ancestor. *Archaeopteryx* took center stage as the no-longer-missing link. In 1982, Harvard neo-Darwinist Ernst Mayr called *Archaeopteryx* "the almost perfect link between reptiles and birds."

But there are too many structural differences between *Archaeopteryx* and modern birds for the latter to be descendants of the former. In 1985 University of Kansas paleontologist Larry Martin wrote: "*Archaeopteryx* is not ancestral of any group of modern birds." Instead, it is "the earliest known member of a totally extinct group of birds." And in 1996 paleontologist Mark Norell, of the American Museum of Natural History in New York, called *Archaeopteryx* "a very important fossil," but added that most paleontologists now believe it is not a direct ancestor of modern birds.

Although there is widespread agreement on this point, there is heated disagreement on another. Which animals might have been the ancestors of *Archaeopteryx*? The controversy involves two different sets of issues: How did flight originate? And how do we go about determining fossil ancestors?

The origin of flight

The evolution of birds from non-flying predecessors would not have been a simple matter, because flight requires extensive modifications to an animal's anatomy and physiology. There are currently two theories of how flight might have originated: the "trees down" theory, and the "ground up" theory. According to the first, the ancestors of birds began their evolutionary journey by leaping from trees, gradually accumulating small adaptations that extended their ability to parachute and glide. According to the second, small animals running after prey on the ground gradually accumulated small adaptations that facilitated their ability to reach and jump. In each theory, the final step was the acquisition of wings and the capacity for true flapping flight.

A major advantage of the "trees down" theory is that gravity presents less of a problem for it than for a "ground up" theory.

It is easier to envisage how animals already in the air might evolve the ability to stay up a little longer, than to envisage how an animal on the ground could evolve the ability to take off. A falling animal would begin by "parachuting," spreading its limbs to break its fall. Small variations that increase its surface area, such as flaps of skin, might give it a slight advantage in the struggle for existence, and future generations might have slightly larger flaps of skin. The second step would be gliding, in which animals with still larger flaps of skin might be able to travel longer distances before coming to earth, like "flying" squirrels. According to the theory, gliding animals eventually achieved true flapping flight.

In the "ground up" theory, birds evolved from animals that ran along the ground chasing prey. Natural selection might favor an ability to run on strong hindlimbs, leaving the forelimbs free to catch the next meal. If selection also favored longer forelimbs to make grasping easier, such animals (according to the theory) might evolve wings and the ability to fly.

An important distinction between the two theories, at least for current controversies over bird evolution, is that they imply very different ancestors for *Archaeopteryx*. The "trees down" theory implies that the ancestors of birds were four-legged reptiles that climbed and jumped from trees, while the "ground up" theory requires two-legged reptiles that ran along the ground and used their forelimbs to catch prey. Four-legged reptiles, of the sort that might have climbed trees, appear in the fossil record well before *Archaeopteryx*. But two-legged reptiles that ran along the ground, and had other features one might expect in an ancestor of *Archaeopteryx*, appear later.

At first glance, the "trees down" theory might seem more plausible. But a relatively new method for analyzing fossils—based on a rigorous application of Darwin's theory—has become quite popular in recent years. The new method is called "cladistics" (from the Greek word meaning "branch"), and it leads to the conclusion that the ancestors of *Archaeopteryx* were two-legged dinosaurs.

Cladistics

Living things are classified into groups based on their similarities. As we saw in the chapter on the tree of life, humans can be grouped with primates, primates with mammals, mammals with vertebrates, and vertebrates with the rest of the animals. This "nested hierarchy" of living things was noticed long before Darwin by Carolus Linnaeus, who devised the modern biological system of classification.

According to Linnaeus, the nested hierarchy reflected the divine plan of creation. According to Darwin, it resulted from the branching-tree pattern of descent from common ancestors. But although Darwin's theory became widely accepted in the 1930s, the Linnaean approach to biological classification was not immediately affected.

By the 1980s, however, most evolutionary biologists were reinterpreting biological classification along Darwinian lines. In 1988 Berkeley biologist Kevin de Queiroz wrote that evolution is "an *axiom* from which systematic methods and concepts are *deduced.*" (emphasis in the original) "Taking evolution as an axiom," de Queiroz continued, "requires that preexisting systematic methods and concepts be reevaluated in its light. Adopting such a perspective should bring the Darwinian Revolution... to fulfillment."

When biological classification is re-interpreted in the light of Darwinian evolution, all groupings become ancestor-descendant sets. Organisms can only be grouped together if they share a common ancestor, and every group includes a common ancestor and all its descendants.

The new perspective, first elaborated by German biologist Willi Hennig in the 1950s, relies for its evidence entirely on homologies. As we saw in the chapter on vertebrate limbs, modern Darwinists define homology as similarity due to common ancestry. Once defined this way, homology cannot be used as evidence for common ancestry without arguing in a circle. In Hennig's approach, organisms are simply assumed to be related by common descent, and their characteristics are then used to infer the points where their lineages diverged into separate branches (hence the name, "cladistics").

In cladistics, character comparisons take precedence over everything else. "The anatomical details or characters," writes paleontologist Pat Shipman, "constitute the evidence, which ultimately adds up to a certainty approaching proof" of evolutionary relationships. Other factors are discounted. For example, physical difficulties inherent in the "ground up" theory of the origin of flight are unimportant; what matters is that birds are anatomically more similar to two-legged running dinosaurs than to four-legged climbing reptiles. To a "cladist" (someone who uses the cladistic method), the debate over the origin of flight is secondary, if not irrelevant.

The order in which animals appear in the fossil record also becomes secondary or irrelevant. If evolutionary relationships are inferred solely on the basis of character comparisons, an animal can be the descendant of another even if the supposed ancestor doesn't appear until millions of years later. The fossil record is simply re-arranged to fit the results of cladistic analysis.

Re-arranging the evidence

Applying cladistics to the evolution of birds leads to the conclusion that the ancestor of *Archaeopteryx* was a two-legged dinosaur.

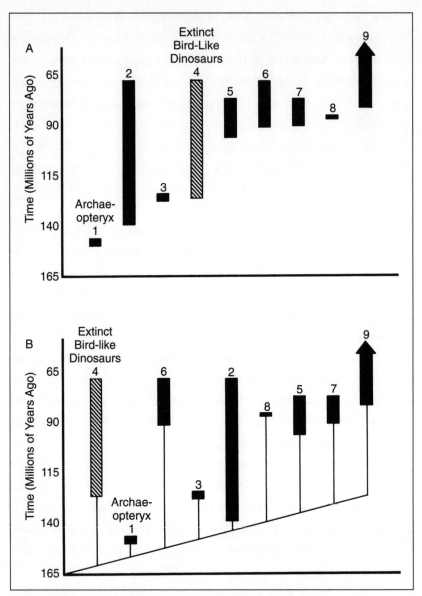

FIGURE 6-2 Cladistic theory and the fossil record.

FIGURE 6-2 **Cladistic theory and the fossil record.**

(A) The actual fossil record of some groups of reptiles and birds, arranged in order of their appearance. The vertical axis represents time, with most recent at the top. The groups are: (1) *Archaeopteryx;* (2-3) two groups of extinct birds; (4) a group of extinct bird-like dinosaurs; (5-8) more groups of extinct birds; and (9) modern birds. (B) Evolutionary relationships among the same groups, according to cladistics. Note the long stretches of hypothetical lineages (thin lines) that are lacking fossil evidence (thick bars).

Indeed, it was the similarity between *Archaeopteryx* and the dinosaur *Compsognathus* that first prompted Huxley to suggest that birds had evolved from dinosaurs. But (as we saw above) that particular dinosaur was discounted as an ancestor of birds because it was the same age as *Archaeopteryx*.

Ironically, once cladistics took over and similarity became the *only* criterion for relationships, paleontologists found that the most likely candidates for the ancestor of *Archaeopteryx* lived tens of millions of years *later*. It was no longer its contemporaneity with *Archaeopteryx* that ruled out *Compsognathus* as an ancestor, but the fact that it didn't have all the right features. According to cladists, the animals with the right features were bird–like dinosaurs that lived in the Cretaceous period, long after *Archaeopteryx* had become extinct. But then, in order to make bird–like dinosaurs the ancestors of birds, the fossil evidence must be re-arranged. (Figure 6-2)

The obvious objection that an animal cannot be older than its ancestor is discounted by assuming that the ancestral form must have been there before its descendant, but its fossil remains cannot be found. In other words, advocates of cladistics cite the imperfection of the geological record—the very same reason

Darwin gave for the troubling absence of transitional forms. As a result, however, the gaps in the fossil record become more pronounced than ever before. Immense stretches of time are left with no fossil evidence to support cladistic phylogenies.

Critics of cladistic methodology argue that the features on which cladists base their analyses may have evolved independently, and don't necessarily point to common ancestry. Critics also argue that although the fossil record is incomplete, it is not as incomplete as cladistic analyses imply. Cladists disagree, and the result has been a raging controversy.

American Museum of Natural History paleontologist Luis Chiappe, a cladist, is untroubled by the implication that birds are descended from dinosaurs that appear to be much younger. "We don't see time as particularly important," Chiappe was quoted as saying in a 1997 *BioScience* article. "We think the fossil record is incomplete." But critic John Ruben, a paleobiologist at Oregon State University, argues that the incompleteness of the fossil record justifies skepticism, not cladistic speculation. "What we ought to be saying is, 'We don't know,'" Ruben was quoted as saying. "So much of this is just hot air."

Whatever the merits of cladistic analysis may be, it has an important consequence for *Archaeopteryx*. It removes the "First Bird" from its iconic status as a missing link, and turns it into just another feathered dinosaur.

Dethroning Archaeopteryx

A cladistic grouping includes a common ancestor and all its descendants, so if birds are descended from dinosaurs then birds *are* dinosaurs. Cladists Lowell Dingus and Timothy Rowe tell their students that birds are "card-carrying" dinosaurs. Although most people think of "dinosaur" as a synonym for obsolescence, Dingus and Rowe claim that the prevalence of birds in the modern world makes dinosaurs "one of Mother Nature's greatest success stories."

The claim that birds are dinosaurs strikes most people— including many biologists—as rather strange. Although it follows from cladistic theory, it defies common sense. Birds and dinosaurs may be similar in some respects, but they are also very different. If birds are dinosaurs, then by the same reasoning humans are fish. As we saw in the chapter on Haeckel's embryos, this sort of "logic" encourages people to see "gill slits" in human embryos that are nothing of the sort.

If cladists are right, then birds are merely feathered dinosaurs. According to Henry Gee, Chief Science Writer for *Nature*, one consequence is the "dethronement" of *Archaeopteryx*. "Once upon a time, *Archaeopteryx* stood alone as the earliest fossil bird. Its uniqueness made it an icon, conferring on it the status of an ancestor," wrote Gee in 1999. But the existence of other bird ancestors (even if their fossils are more recent) "shows that *Archaeopteryx* is just another dinosaur with feathers."

But if *Archaeopteryx* is no longer the missing link, what is? Ironically, the cladistic revolution has resurrected the search for transitional forms that *Archaeopteryx* was supposed to have ended. Now every few months some paleontologist announces the discovery of another "missing link," as though the First Bird had never been found. *Archaeopteryx*, the bird in hand, has been abandoned for two in the bush. One recent consequence has been the most embarrassing fossil fraud since Piltdown.

The "Piltdown bird"

In 1912 amateur geologist Charles Dawson and the British Museum announced the discovery near Piltdown, England, of a missing link between apes and humans. The specimen lay in the British Museum until it was exposed as a fake in 1953. Someone had combined an ancient human skull with the lower jaw of a modern orangutan, modified to look like part of the same individual. "Piltdown man" (to whom we shall return in Chapter 11) remains the most famous fossil fraud in the history of science.

In 1999 amateur dinosaur enthusiast Stephen Czerkas and the National Geographic Society announced that a fossil purchased for $80,000 at an Arizona mineral show was "the missing link between terrestrial dinosaurs and birds that could actually fly." The fossil, which was apparently smuggled out of China, had the forelimbs of a primitive bird and the tail of a dinosaur. Czerkas named it *Archaeoraptor*.

In November 1999 *National Geographic* magazine featured *Archaeoraptor* in an article entitled "Feathers for T. rex?" Christopher Sloan, the article's author, claimed that we can now say that birds are dinosaurs "just as confidently as we say that humans are mammals," and that feathered dinosaurs preceded the first bird. The article featured a drawing of a baby *Tyrannosaurus* with feathers—hence its title. It also included a picture of the *Archaeoraptor* fossil, explaining that its combination of "advanced and primitive features is exactly what scientists would expect to find in dinosaurs experimenting with flight."

It turns out that *Archaeoraptor* had exactly the features scientists were expecting to find because a clever forger had fabricated it that way, knowing it would bring big bucks in the international fossil market. The fabrication was discovered by Chinese paleontologist Xu Xing, who proved that the specimen consisted of a dinosaur tail glued to the body of a primitive bird.

Storrs Olson, curator of birds at the Smithsonian Institution in Washington, D.C., fired off an angry letter to Peter Raven, Secretary of the National Geographic Society. Olson blasted the Society for allying itself with "a cadre of zealous scientists" who have become "outspoken and highly biased proselytizers of the faith" that birds evolved from dinosaurs. "Truth and careful scientific weighing of evidence have been among the first casualties in their program," wrote Olson, "which is fast becoming one of the grander scientific hoaxes of our age."

National Geographic posted a partial retraction January 21, 2000, on its Internet web site. Nevertheless, the magazine was severely criticized in February by *Nature* for "naively and hastily publishing an article—described as 'sensationalistic, unsubstantiated, tabloid journalism' by a leading paleontologist—sprinkled with dubious assertions."

The incident was acutely embarrassing for *National Geographic,* which attempted to lay it to rest by publishing a letter about the fraud from Xu Xing in March 2000. Meanwhile, the magazine's editor protested the *Nature* editorial, claiming that "pertinent information concerning the integrity of the specimen" had been withheld from *National Geographic* and from the scientists it had paid to study the fossil.

Charges and counter-charges continue to fly. Some people involved in the scandal blame it on the international trade in smuggled fossils, while others blame it on shoddy journalism. But the real culprit seems to be the cladists' desire to prove their theory. Just as the need for a missing link between apes and humans led to Piltdown man, so the need for a missing link between dinosaurs and birds paved the way for the "Piltdown bird." Lost in the hubbub was the fact that even if *Archaeoraptor* had been genuine, it was tens of millions of years younger than

Archaeopteryx, and thus would have failed to plug the gap left in the fossil record by cladistic methodology.

In April 2000 Czerkas and prominent cladists—together with some of their critics—gathered in Fort Lauderdale, Florida, for a Symposium on Dinosaur Bird Evolution. I attended, as well, to listen in on the controversy. Although some had feared that the embarrassing *Archaeoraptor* episode would dominate the conference, the fraud was largely ignored. In its place, cladists presented their new star, advertised to be the best missing link yet.

Feathers for Bambiraptor

The new discovery that upstaged the *Archaeoraptor* fraud was *Bambiraptor*, originally discovered by a Montana family in 1993 and turned over to professional paleontologists in 1995. The animal's body was about the size of a chicken, but its long tail made it about three feet long. With sharp teeth and claws, it resembled a small *Velociraptor*—the ruthless predator made famous in the closing scenes of the movie "Jurassic Park."

The original skeleton of *Bambiraptor*—reconstructed in a lifelike pose and protected by thick Plexiglas—was proudly displayed at the conference. (Figure 6-3) The fossil had been found in Upper Cretaceous rocks, meaning that it was about 75 million years *younger* than *Archaeopteryx*. But cladistic analysis showed that it had many of the skeletal features predicted to have existed in the ancestor of *Archaeopteryx*. In fact, paleontologists who examined it proclaimed it to be "the most bird-like dinosaur yet discovered" and a "remarkable missing link between birds and dinosaurs."

FIGURE 6-3 *Bambiraptor.*

Reconstructed skeleton displayed at the April 2000 Florida Symposium on Dinosaur Bird Evolution.

Brian Cooley, who specializes in reconstructing dinosaurs from fossil skeletons, had reconstructed *Bambiraptor* for the conference exhibit. He explained to the participants that he set out to make *Bambiraptor* as bird-like as possible, given its supposed position between dinosaurs and birds. He reconstructed the muscles using bird anatomy as his guide, and he placed the eyes in a bird-like orientation, using the same artificial eyes taxidermists put in stuffed eagles. Guessing that *Bambiraptor* must have been covered with "scruffy" feathers, Cooley added them to his reconstruction. (Figure 6-4)

Every conference attendee was given a copy of the article containing the official scientific description of *Bambiraptor,* published just three weeks earlier. The first published report of a newly discovered fossil species is supposed to conform to the highest scientific standards, describing the "type" specimen with scrupulous attention to accuracy. The official description of *Bambiraptor* contains several drawings of the reconstructed animal, two of which show hair-like projections on the body and feathers on the forelimbs.

But nothing remotely resembling feathers was found with the fossil. The hair-like projections and the feathers are imaginary. Because cladistic theory says they should be there, they were included in the scientific description of the fossil. The only indication in the article that the projections and feathers are not real is a figure caption that includes the line: "Reconstruction showing conceptual integumentary structures." I was surprised. Ordinarily, one might expect something in reasonably plain English, such as: "Hair-like projections and feathers were not found with the fossil, but have been added here based on theoretical considerations." Under the circumstances, the article seemed better designed to obscure the truth than to report it.

FIGURE 6-4 Feathered *Bambiraptor.*

Reconstructed animal displayed at the April 2000 Florida Symposium on Dinosaur Bird Evolution showing "conceptual integumentary structures."

There were several outspoken critics of the dino-bird theory at the Florida symposium. One was University of North Carolina ornithologist Alan Feduccia, who has predicted that the dino-bird theory will turn out to be "the greatest embarrassment of paleontology of the 20th century." Another was Larry Martin, who has said that if he had to defend the dino-bird theory, "I'd be embarrassed every time I had to get up and talk about it." And Storrs Olson ruffled some dino-feathers by passing out buttons that proclaimed "Birds are NOT dinosaurs."

But the dino-bird enthusiasts at the symposium outnumbered their critics, and they were undeterred from dressing up *Bambiraptor* in imaginary feathers. Not being a cladist myself, I found this rather funny. As a molecular biologist, however, I found something else even funnier.

Turkey DNA from Triceratops?

On the second day of the symposium, William Garstka reported that he and a team of molecular biologists from Alabama had extracted DNA from the fossil bones of a 65-million-year-old dinosaur. Although evidence from other studies suggests that DNA older than about a million years cannot yield any useful sequence information, Garstka and his colleagues amplified and sequenced the DNA, compared it with known DNA from other animals, and found that it was most similar to bird DNA. They concluded that they had found "the first direct genetic evidence to indicate that birds represent the closest living relatives of the dinosaurs." Their conclusion was reported the following week by Constance Holden in *Science*.

The details of the discovery, however, are revealing. First, the dinosaur from which Garstka and his colleagues allegedly recovered the DNA was a *Triceratops*. According to paleontologists, there are two main branches in the dinosaur family tree. One branch included the three-horned rhinoceros-like *Triceratops* which millions of people have seen in museum exhibits and movies. But birds are thought to have evolved from the other branch. So according to evolutionary biologists, *Triceratops* and modern birds are not closely related, their ancestors having gone their separate ways almost 250 million years ago.

Even more revealing, however, was that the DNA Garstka and his colleagues found was *100 percent identical to the DNA of living turkeys*. Not 99 percent, not 99.9 percent, but 100 percent. Not even DNA obtained from other birds is 100 percent identical to turkey DNA (the next closest match in their study was 94.5 percent, with another species of bird). In other words, the DNA that had supposedly been extracted from the *Triceratops* bone was not just similar to turkey DNA—it *was* turkey DNA. Garstka said he and his colleagues considered the possibility that someone had been eating a turkey sandwich nearby, but they were unable to confirm that.

At first, when Garstka presented his findings I thought it was an April Fool's joke—but it was already April 8. Then I looked around to see whether anyone was laughing—but no one was, at least not openly. When I returned home the next day and told my wife the story, she said it reminded her of a child who botches an attempt to stay home from school. When the child's mother puts a thermometer in his mouth, he holds it up to a light bulb to drive the temperature up, but he holds it there too long. When mom returns and sees that his temperature is 130 degrees, she sends him packing. The moral of the story is: If you're going to fake something, don't make it so obvious. The DNA from *Triceratops* might not have been so funny if it hadn't been 100 percent identical to turkey DNA.

In all fairness, Garstka admitted that he was skeptical of the results—not only because of the possible turkey sandwich, but also because nobody thinks birds are descended from *Triceratops*. Of course, strange things happen, but the "extraction" of turkey DNA from *Triceratops* had all the earmarks of a hoax—perhaps a hoax perpetrated on Garstka and his colleagues by someone else.

The incident convinced me that some people are so eager to believe that birds evolved from dinosaurs that they are willing to accept almost any evidence that appears to support their view, no matter how far-fetched. The other side of the coin, of course, is an unwillingness to give a fair hearing to critics of their view. And the other side of the coin was well represented by the speaker who had preceded Garstka on the platform.

The "cracked kettle" approach to doing science

Just before Garstka spoke, Berkeley paleontologist Kevin Padian had blasted critics of the dino-bird theory for being unscientific. Padian explained that, as President of the National Center for Science Education, he spends a lot of time telling people what science is and what it isn't. (The National Center for Science Education—despite its neutral-sounding title—is a pro-Darwin advocacy group that discourages public schools from exposing students to controversies over evolution.) Padian emphasized that science is about testing hypotheses with evidence. If we can't test an idea, it isn't necessarily false, but it isn't scientific.

Padian called critics of the dino-bird hypothesis unscientific because (he claimed) they offer no empirically testable alternative hypotheses. The evidence the critics cite for their hypotheses, he claimed, is based on the "selective interpretation of isolated observations," rather than on a method (cladistics) that is "fully accepted by the scientific community." Although "science is not a vote," the cladistic method is endorsed by the National Science Foundation, major peer-reviewed scientific journals, and "the majority of experts." Therefore, criticisms of the dino-bird hypothesis "ceased to be science more than a decade ago," and the "controversy is dead."

Needless to say, the announcement that the controversy was dead failed to persuade the critics in the audience. But the most amazing thing about Padian's lecture was its stunning display of non-sequiturs. In fact, it reminded me of an old lawyers' joke.

According to the joke, Jones sues Smith for borrowing his kettle and returning it with a crack in it. Smith's lawyer defends him as follows:

1. Smith never borrowed the kettle.
2. When Smith returned the kettle, it wasn't cracked.
3. The kettle was already cracked when Smith borrowed it.
4. There is no kettle.

Of course, Padian was not trying to be funny, and it may seem unkind to compare his talk to an old lawyers' joke. But consider the following summary of his argument:

1. In the controversy over bird origins, critics of the dinosaur hypothesis have not proposed any alternative hypotheses that can be tested by evidence.
2. The evidence on which the critics base their alternative hypotheses is selectively interpreted.
3. Although science is not a vote, the majority of the scientific community rejects the critics' methodology regardless of their evidence.
4. There is no controversy.

Now, Kevin Padian takes his work seriously. So do the people who paid $80,000 for the Piltdown bird, the paleontologists who put imaginary feathers on *Bambiraptor*, and the molecular biologists who reported finding turkey DNA in *Triceratops*. But as I left the Florida symposium I couldn't help chuckling. So much of what I had seen and heard seemed downright silly. In fact, if I had been an artist instead of a biologist, I might have sketched some cartoons, with captions such as these:

"Dino-bird enthusiasts find fossils made to order."
"Cladistic mob tars and feathers defenseless dinosaur."
"Turkey sandwich proves birds evolved from *Triceratops.*"
"Old lawyers' joke becomes new scientific method."

This isn't science. This isn't even myth. This is comic relief. But after we've had a good laugh we need to ask ourselves: Whatever happened to *Archaeopteryx*?

Whatever happened to Archaeopteryx?

Some biology textbooks continue to present *Archaeopteryx* as the classic example of a missing link. Mader's 1998 *Biology* calls it "a transitional link between reptiles and birds," and William Schraer and Herbert Stoltze's 1999 *Biology: The Study of Life* tells students that "many scientists believe it represents an evolutionary link between reptiles and birds."

But both sides in the current controversy over bird origins agree that modern birds are probably not descended from *Archaeopteryx*. And although the two factions disagree about the ancestry of *Archaeopteryx,* neither one has really solved the problem. Following the logic of Darwin's theory to sometimes silly extremes, cladists insist that the ancestors of *Archaeopteryx* were bird-like dinosaurs that do not appear in the fossil record until tens of millions of years later. Their critics look to animals that clearly lived earlier, but have not yet found one similar enough to *Archaeopteryx* to be a good candidate. As a result, *both* sides are still looking for the missing link.

Isn't it ironic that *Archaeopteryx*, which more than any other fossil persuaded people of Darwin's theory in the first place, has been dethroned largely by cladists, who more than any other biologists have taken Darwin's theory to its logical extreme? The world's most beautiful fossil, the specimen Ernst Mayr called "the almost perfect link between reptiles and birds," has been quietly shelved, and the search for missing links continues as though *Archaeopteryx* had never been found.

CHAPTER 7

Peppered Moths

Darwin was convinced that in the course of evolution "Natural Selection has been the most important, but not the exclusive, means of modification," but he had no direct evidence of natural selection. There was plenty of evidence that plants and animals vary, and that they struggle for survival. It was reasonable to conclude, by analogy with domestic breeding, that organisms with the most advantageous variations would survive and pass them on to their offspring. But no one had actually documented this process in the wild. The best Darwin could do in *The Origin of Species* was "give one or two imaginary illustrations."

It wasn't until 1898 that something approaching direct evidence for natural selection was provided by Brown University biologist Hermon Bumpus. After a severe snowstorm in Providence, Rhode Island, Bumpus had found a large number of English sparrows close to death. He took over a hundred of them back to his laboratory, where almost half died. When he measured and compared the living and the dead, he found that the survivors tended to be males that were shorter and lighter. Apparently, the blizzard had selected against females and larger

FIGURE 7-1   Peppered moths resting on tree trunks.

FIGURE 7-1 **Peppered moths resting on tree trunks.**

(Top) Two moths (one typical and one melanic) resting on the dark bark of an oak tree in a polluted woodland. (Bottom) Typical and melanic moths resting on the lichen-covered trunk of an oak tree in an unpolluted woodland. Note the striking differences in camouflage.

males; but it was not clear why, so the actual reason for the selection remained elusive. For several decades, though, Bumpus's work was the closest biologists had come to observing natural selection directly.

But even as Bumpus was measuring his sparrows, British scientists were noticing another phenomenon that would eventually become the classic textbook example of natural selection in action. Most peppered moths were light-colored in the early part of the nineteenth century, but during the industrial revolution in Britain the moth populations near heavily polluted cities became predominantly "melanic," or dark-colored. The phenomenon was called "industrial melanism," but its causes remained a matter of speculation until the early 1950s, when British physician and biologist Bernard Kettlewell performed some experiments which made him famous. Kettlewell's experiments suggested that predatory birds ate light-colored moths when they became more conspicuous on pollution-darkened tree trunks, leaving the dark-colored variety to survive and reproduce. Industrial melanism in peppered moths appeared to be a case of natural selection.

Most introductory biology textbooks now illustrate this classical story of natural selection with photographs of the two varieties of peppered moth resting on light- and dark-colored tree trunks. (Figure 7-1) What the textbooks don't explain, however, is that biologists have known since the 1980s that the classical story has some serious flaws. The most serious is that peppered moths in the wild don't even rest on tree trunks. The textbook photographs, it turns out, have been staged.

Industrial melanism

The peppered moth, *Biston betularia*, comes in various shades of gray. One hundred and fifty years ago, most peppered moths were "typical" forms, which have predominantly light gray scales with a few black scales scattered among them (hence the name, "peppered"). As early as 1811, however, the species also included some coal-black "melanic" forms. During the industrial revolution, the proportion of melanic forms increased, and by the turn of the century more than 90% of the peppered moths near the industrial city of Manchester, England, were melanic.

A similar increase in melanic forms was reported in many other species of moths, ladybird beetles, and even some birds. It was also reported near other industrial cities such as Birmingham and Liverpool. Obviously, this was not an isolated phenomenon, and the name "industrial melanism" was used to denote all its manifestations.

In 1896 British biologist J. W. Tutt suggested that industrial melanism in peppered moths might be due to differences in camouflage. Tutt theorized that in unpolluted woodlands, typicals are well camouflaged against the light-colored lichens that grow on tree trunks; but in woodlands where industrial pollution has killed the lichens and darkened the tree trunks, melanics are better camouflaged. Since predatory birds could be expected to find and eat the more conspicuous moths, the proportion of melanic forms would increase as a result of natural selection.

In the 1920s another British biologist, J. W. Heslop Harrison, rejected Tutt's theory and proposed that melanism was induced directly by airborne industrial pollutants. Although he did not work on *Biston betularia*, Harrison reported that melanism could be produced in several other moth species if their larvae were fed on leaves contaminated with metallic salts. Critics were unable to reproduce Harrison's results, however, and pointed out that some of the species Harrison tested did not exhibit industrial melanism in the wild.

There was a theoretical problem with Harrison's work, as well. If melanism could be induced it meant that the organism acquired it after birth. But there was also clear evidence that melanism was inherited, so Harrison's view implied that acquired characteristics could later be inherited. According to neo-Darwinian theory, however, the inheritance of acquired characteristics was impossible; all new heritable variations arose from genetic changes such as mutation.

As neo-Darwinism rose in popularity, the influence of Harrison's ideas declined, and most biologists adopted the theory that industrial melanism in peppered moths was due to natural selection. It wasn't until the 1950s, however, that British physician and biologist Bernard Kettlewell set out to test the theory empirically.

Kettlewell's experiments

Like Tutt, Kettlewell believed that melanic moths increased in number because of camouflage and predatory birds, and he performed several experiments to test the theory. First, to determine whether birds preyed on peppered moths at all, he released some moths into an aviary containing a pair of nesting birds and their young. Then he watched through binoculars as the moths settled onto various resting sites and were eaten by the birds.

Having established that birds actually prey on peppered moths, Kettlewell released some moths onto tree trunks in a polluted woodland near Birmingham, England. He watched through binoculars as the moths settled on nearby tree trunks, and noted that melanics were much less conspicuous than typicals, as judged by the human eye. He also observed that birds took the conspicuous moths more readily than inconspicuous ones.

Kettlewell then marked several hundred peppered moths, typicals as well as melanics, with tiny dots of paint on the underside of their wings, and released them during the day onto nearby tree trunks in the polluted Birmingham woodland. On the following nights he set out traps to recapture as many as he could. Of the 447 marked melanics he released, he recaptured 123, while of 137 marked typicals he recaptured only 18. In other words, he recaptured 27.5 percent of the melanics, but only 13.0 percent of the typicals. Kettlewell concluded that a much higher proportion of melanics had survived predation, and that "birds act as selective agents, as postulated by evolutionary theory."

Two years later, Kettlewell repeated the same procedure in an unpolluted woodland in Dorset, England. Once again he released moths onto nearby tree trunks. As expected, melanic moths were much more conspicuous than typicals on the lichen-covered Dorset trees, and thus more readily taken by predatory birds. Famed animal behaviorist Niko Tinbergen accompanied Kettlewell and made movies of birds picking the moths off tree trunks.

Then Kettlewell repeated his mark-release-recapture experiment by marking and releasing hundreds of moths onto the unpolluted tree trunks, and recapturing as many as he could on subsequent nights. Of the 496 marked typicals he released, he recaptured 62 (12.5 percent), but of the 473 marked melanics he recaptured only 30 (6.3 percent), so the two-to-one ratio he had obtained in Birmingham was completely reversed. Kettlewell concluded that typicals enjoyed a selective advantage in Dorset because their superior camouflage on lichen-covered tree trunks improved their chances of surviving hungry birds.

Darwin's missing evidence?

Kettlewell called industrial melanism in peppered moths "the most striking evolutionary change ever actually witnessed in any organism." Since his experiments seemed to provide empirical confirmation of natural selection, Kettlewell dubbed his results "Darwin's missing evidence" in an article written for *Scientific American*.

Following the passage of anti-pollution legislation in the 1950's, industrial melanism began to decline. The percentage of melanic peppered moths west of Liverpool dropped slightly between 1959 and 1962, and a decade later the reversal of industrial melanism was well under way. Field studies in the 1960s and 1970s showed that the proportion of typicals rose as pollution decreased, consistent with the theory that industrial melanism in peppered moths was due to camouflage and predatory birds.

In 1975 British geneticist P. M. Sheppard called the phenomenon "the most spectacular evolutionary change ever witnessed and recorded by man, with the possible exception of some examples of pesticide resistance," and famed evolutionary biologist Sewall Wright called it "the clearest case in which a conspicuous evolutionary process has actually been observed."

A critic of Darwin's theory might object that this "most spectacular evolutionary change ever witnessed" falls far short of providing a sufficient mechanism for evolution. After all, the only

thing that happened was a change in the proportion of two varieties of a pre-existing species of moth. Although the change was
dramatic, it was no more impressive than the changes domestic
breeders have been producing for centuries.

But in the 1950s, Kettlewell's evidence for a "conspicuous
evolutionary process" was the best available. Industrial melanism
in peppered moths—and Kettlewell's explanation of its cause—
became the classic textbook example of natural selection in
action. Yet while peppered moths were being transformed into
icons of evolution, discrepancies began to appear that eventually
cast serious doubt on the validity of Kettlewell's experiments.

Problems with the evidence

When biologists looked beyond Birmingham and Dorset, where
Kettlewell had conducted his experiments, they found some
discrepancies between Kettlewell's explanation and the actual
geographical distribution of melanic moths. For example, if
melanic moths in polluted woodlands enjoyed as much of a
selective advantage as Kettlewell's experiments seemed to indicate, then they should have completely replaced typicals in heavily polluted areas such as Manchester. This never happened,
however, suggesting that factors other than camouflage and
predatory birds must be involved.

Some other distribution features were inconsistent with Kettlewell's explanation, as well. In rural Wales, the frequency of
melanics was higher than expected, prompting Liverpool biologist Jim Bishop to conclude in 1972 that "as yet unknown factors" were involved. In rural East Anglia (Figure 7-2, B), where
there was little industrial pollution and typicals seemed better
camouflaged, melanics reached a frequency of 80 percent,

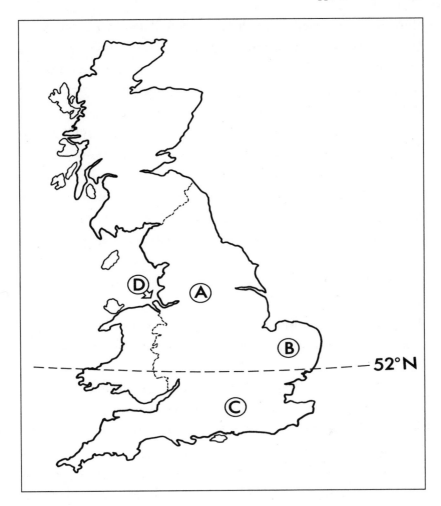

FIGURE 7-2 Discrepancies in peppered moth distribution.

Locations in the United Kingdom of some discrepancies that didn't fit the classical story. (A) Manchester, where the proportion of melanics was never as
high as theory predicted; (B) East Anglia, where melanism was high despite
lichen-covered tree trunks; (C) south of latitude 52ºN, where melanism
increased after the introduction of pollution control; (D) the Wirral Peninsula,
where melanism began decreasing before lichens returned to the trees.

prompting two other biologists to conclude in 1975 "that either the predation experiments and tests of conspicuousness to humans are misleading, or some factor or factors in addition to selective predation are responsible for maintaining the high melanic frequencies."

On the other hand, melanics in south Wales seemed better camouflaged than typicals, yet they comprised only about 20 percent of the population. Compiling data from 165 separate sites in Britain, R. C. Steward found a correlation between melanism and the concentration of sulfur dioxide (an airborne pollutant) north—but not south—of latitude 52°N. (Figure 7-2, C) Steward concluded that "in the south of Britain non-industrial factors may be of greater importance" than camouflage and bird predation.

After the passage of anti-pollution legislation, the proportion of melanics north of London decreased as expected, but inexplicably increased in the south. Theoretical models could account for the discrepancies only by invoking migration and unknown "non-visual selective factors." Whatever had caused industrial melanism, it was clearly more than camouflage and bird predation.

In other words, Kettlewell's explanation had been too simple. Not surprisingly, the actual situation was turning out to be more complicated. And geographical discrepancies were not the only complication. In the 1970s and 1980s biologists realized that melanism was not well correlated with changes in lichens.

The exaggerated role of lichens

If the rise of melanism was due to the darkening of tree trunks following the loss of their lichen cover from pollution, then a reduction in pollution should bring lichens back to the trees and lead to a reversal of industrial melanism. The reversal occurred, but it happened without the predicted return of the lichens.

In the 1970s Kettlewell himself noted that melanism began declining on the Wirral Peninsula before lichens reappeared. (Figure 7-2, D) When David Lees and his colleagues surveyed melanism in peppered moths at 104 sites throughout Britain, they found a lack of correlation with lichen cover which they considered "surprising in view of the results of Kettlewell's selection experiments."

In the early 1980s Cyril Clarke and his colleagues found "a reasonable correlation" in the U.K. between the decline in melanism and decrease in sulfur dioxide pollution, but were surprised to note "that throughout this time the appearance of the trees in Wirral does not seem to have changed appreciably." American biologist Bruce Grant and Cambridge biologist Rory Howlett noted in 1988 that if the rise of industrial melanism had originally been due to the demise of lichens on trees, then "the prediction is that lichens should precede the recovery of the typical morph as the common form. That is, the hiding places should recover before the hider." But their field work showed that "this is clearly not the case in at least two regions where the recovery of typicals has been especially well documented in the virtual absence of these lichens: on the Wirral... and in East Anglia."

While melanism was rising and falling in the United Kingdom, it was doing the same in the United States. The first American melanic peppered moth was reported near Philadelphia in 1906, and the proportion of melanics rose rapidly thereafter. By 1960 the proportion of melanics in southeastern Michigan was over

90 percent. When pollution-control measures were introduced, melanism underwent the same sort of reversal that was observed in the United Kingdom, and by 1995 the frequency of melanics in southeastern Michigan had dropped to less than 20 percent.

But the decline of melanism in the United States was not correlated with changes in the lichen cover on tree trunks. In Michigan, for example, it "occurred in the absence of perceptible changes" in local lichen cover, prompting Grant and his colleagues to conclude that "the role of lichens has been inappropriately emphasized in chronicles about the evolution of melanism in peppered moths."

So in the United States as well as in the United Kingdom, melanism declined before lichens returned to the trees. Apparently, the presence or absence of lichens was not as important as Kettlewell had thought. The discrepancy was significant, and pointed to a deeper problem. It turns out that Kettlewell's experiments, and most of the other experiments performed in the 1960s and 1970s, had not used the natural resting places of peppered moths.

Peppered moths don't rest on tree trunks

In most of Kettlewell's experiments, moths were released and observed during the day. In only one experiment (June 18, 1955) did Kettlewell release moths at night, just before sunrise. He immediately abandoned this approach because of the practical difficulties it entailed, such as having to warm the cold moths beforehand on the engine of his car. But peppered moths are night-fliers, and normally find resting places on trees before dawn. The moths Kettlewell released in the daytime remained exposed, and became easy targets for predatory birds. Regarding his release methods, Kettlewell wrote: "I admit that, under their own choice, many would have taken up position higher in the trees." He assumed, however, that he could disregard the artificiality of his technique.

Before the 1980s most investigators shared Kettlewell's assumption, and many of them found it convenient to conduct predation experiments using *dead* specimens glued or pinned to tree trunks. Kettlewell himself considered this a bad idea, and even some biologists who used dead moths suspected that the technique was unsatisfactory. For example, Jim Bishop and Laurence Cook conducted predation experiments using dead moths glued to trees; but they noted discrepancies in their results which "may indicate that we are not correctly assessing the true nature of the resting sites of living moths when we are conducting experiments with dead ones."

Since 1980, however, evidence has accumulated showing that *peppered moths do not normally rest on tree trunks*. Finnish zoologist Kauri Mikkola reported an experiment in 1984 in which he used caged moths to assess normal resting places. Mikkola observed that "the normal resting place of the Peppered Moth is beneath small, more or less horizontal branches (but not on narrow twigs), probably high up in the canopies, and the species probably only exceptionally rests on tree trunks." He noted that "night-active moths, released in an illumination bright enough for the human eye, may well choose their resting sites as soon as possible and most probably atypically."

Although Mikkola used caged moths, data on wild moths supported his conclusion. In twenty-five years of field work, Cyril Clarke and his colleagues found only one peppered moth naturally perched on a tree trunk; they concluded that they knew primarily "where the moths do *not* spend the day." When Rory

Howlett and Michael Majerus studied the natural resting sites of peppered moths in various parts of England, they found that Mikkola's observations on caged moths were valid for wild moths, as well. "It seems certain that most *B. betularia* rest where they are hidden," they concluded, and that "exposed areas of tree trunks are not an important resting site for any form of *B. betularia*." In a separate study reported in 1987, British biologists Tony Liebert and Paul Brakefield confirmed Mikkola's observations that "the species rests predominantly on branches.... Many moths will rest underneath, or on the side of, narrow branches in the canopy."

In a 1998 book on industrial melanism, Michael Majerus defended the classical story but criticized the "artificiality" of much of the work on peppered moths, noting that in most predation experiments they were "positioned on vertical tree trunks, despite the fact that they rarely chose such surfaces to rest upon in the wild." But if peppered moths don't rest on tree trunks, where did all those photographs come from?

Staged photographs

Pictures of peppered moths on tree trunks must be staged. Some are made using dead specimens that are glued or pinned to the trunk, while others use live specimens that are manually placed in desired positions. Since peppered moths are quite torpid in daylight, they remain where they are put.

Manually positioned moths have also been used to make television nature documentaries. University of Massachusetts biologist Theodore Sargent told a *Washington Times* reporter in 1999 that he once glued some dead specimens on a tree trunk for a TV documentary about peppered moths.

Staged photos may have been reasonable when biologists thought they were simulating the normal resting-places of peppered moths. By the late 1980s, however, the practice should have stopped. Yet according to Sargent, a lot of faked photographs have been made since then.

Defenders of the classical story typically argue that, despite being staged, the photographs illustrate the true cause of melanism. The problem is that it is precisely the cause of melanism that is in dispute.

Doubts about the classical story

When birds preyed on Kettlewell's moths, the moths were not in their natural hiding places. This one fact casts serious doubt on the validity of his experiments. In the mid-1980s, Italian biologists Giuseppe Sermonti and Paola Catastini criticized Kettlewell's daytime releases and concluded that his experiments "do not prove in any acceptable way, according to the current scientific standard, the process he maintains to have experimentally demonstrated." Sermonti and Catastini concluded that "the evidence Darwin lacked, Kettlewell lacked as well."

With Kettlewell's evidence impeached, some biologists now argue that Heslop Harrison's hypothesis of direct induction by pollutants deserves another look. According to Japanese biologist Atuhiro Sibatani, "the story of industrial melanism must be shelved, at least for the time being, as a paradigm of neo-Darwinian evolution," and Harrison's work should be re-examined. Sibatani maintains that an inordinate devotion to neo-Darwinian theory led to a "sheer dismissal" of the induction hypothesis and a "too optimistic acceptance of the shaky evidence for the natural selection model of industrial melanism."

Most biologists, however—even critics of Kettlewell's work—believe that the principal cause of industrial melanism was natural selection rather than induction. For them, the dispute is over what selective factors were involved. In 1998 American biologist Theodore Sargent and his New Zealander colleagues Craig Millar and David Lambert wrote: "We feel certain that this phenomenon is a product of selection," though the intuitive appeal of Kettlewell's explanation "may have blinded us to the role that other selective factors might be playing in the melanism story." Sargent and his colleagues listed several factors, including possible differences in the tolerance of larvae to pollutants, or in the moths' vulnerability to parasites, and concluded that "the complex of factors that might play a role in the increase (or decrease) of melanism in moths has barely been tapped."

It is interesting to note that other selective factors were responsible for industrial melanism in ladybird beetles. Birds find the beetles extremely distasteful, and will not eat them, so camouflage and bird predation played no role. Melanic ladybird beetles are thought to be more fit in a smoky environment because they are better able to absorb solar radiation—a phenomenon known as "thermal melanism." Although no one maintains that thermal melanism was at work in peppered moths, this example shows that industrial melanism may have other causes.

The need to consider other causes does not mean that camouflage and bird predation are irrelevant. In fact, they may still be the most important factors in the rise and fall of industrial melanism in peppered moths. British biologists Michael Majerus and Laurence Cook cite various other observations in defense of the classical story, and continue to defend it, though they also acknowledge that further work is needed.

In any case, it is clear that the compelling evidence for natural selection that biologists once thought they had in peppered moths no longer exists. As Sargent and his colleagues wrote in 1998, "the 'classical' explanation may be true, in whole or in part. We contend, however, that there is little persuasive evidence, in the form of rigorous and replicated observations and experiments, to support this explanation at the present time." It seems that "Darwin's missing evidence" for natural selection—at least in peppered moths—is still missing.

Nevertheless, controversy over the classical story continues, and it highlights an important question: What does it take to demonstrate natural selection scientifically?

Science or alchemy?

In 1998 University of Chicago evolutionary biologist Jerry Coyne wrote a review in *Nature* of Michael Majerus's book, *Melanism: Evolution in Action*. As we have seen, Majerus defended the classical story, but he also acknowledged the problems with it. And the problems were enough to convince Coyne that the story is in serious trouble. "From time to time," Coyne wrote, "evolutionists re-examine a classic experimental study and find, to their horror, that it is flawed or downright wrong." According to Coyne, the fact that peppered moths do not rest on tree trunks "alone invalidates Kettlewell's release-and-recapture experiments, as moths were released by placing them directly onto tree trunks."

After he went back to Kettlewell's original papers and "unearthed additional problems," Coyne concluded that this "prize horse in our stable of examples" of evolution "is in bad shape, and, while not yet ready for the glue factory, needs serious attention." Especially in need of attention, argued Coyne, are the selective factors responsible for industrial melanism. It is not enough merely to claim that a phenomenon is due to natural selection. It is also necessary "to unravel the forces changing a character. We must stop pretending that we understand the course of natural selection" just because we know that one trait is more fit than another.

But College of William and Mary biologist Bruce Grant rushed to the defense of the classical story. While acknowledging that things are more complicated than they appear in textbooks, Grant insists that "the evidence in support of the basic story is overwhelming." The evidence Grant cites, however, is surprisingly thin. He admits that "we still don't know the natural hiding places of peppered moths," he agrees that "the greatest weakness of Kettlewell's mark-release-recapture experiments is that he released the moths during daylight hours," and he repeats his own finding that most accounts of peppered moths "place too much attention on the importance of lichens."

Yet Grant claims that Kettlewell's results are valid anyway. There is "indisputable evidence for natural selection," he argues, because "even if all of the experiments relating to melanism in peppered moths were jettisoned, we would still possess the most massive data set on record" for a conspicuous evolutionary change. Grant concludes that "no other evolutionary force can explain the direction, velocity, and the magnitude of the changes except natural selection."

Evidence for industrial melanism, however, is not necessarily evidence for natural selection, and it is certainly not evidence that the selective agents were predatory birds. As we saw above, melanic forms might survive better in a polluted environment for any number of reasons, and even biologists who defend the general outline of the classical story acknowledge that "non-visual selective factors" must also have been involved. No one doubts that a change in the proportion of the two varieties of peppered moth occurred. But what caused it?

In 1986 evolutionary biologist John Endler wrote a book entitled *Natural Selection in the Wild*, now acknowledged to be a classic in the field. At the time, Endler was unaware of the problems being unearthed in the peppered moth story, so he listed it as one of the few cases in which the cause of natural selection was known. But he also declared that "the time has passed for 'quick and dirty' studies of natural selection." Although most researchers are "satisfied in demonstrating merely that natural selection occurred," Endler wrote, "this is equivalent to demonstrating a chemical reaction, and then not investigating its causes and mechanisms. A strong demonstration of natural selection combined with a lack of knowledge of its reasons and mechanisms is no better than alchemy."

Industrial melanism in peppered moths shows that the relative proportions of two pre-existing varieties can change dramatically. This change may have been due to natural selection, as most biologists familiar with the story believe. But Kettlewell's evidence for natural selection is flawed, and the actual causes of the change remain hypothetical. As a scientific demonstration of natural selection—as "Darwin's missing evidence"—industrial melanism in peppered moths is no better than alchemy.

Open almost any biology textbook dealing with evolution, however, and you'll find the peppered moth presented as a classical demonstration of natural selection in action—complete with faked photos of moths on tree trunks. This is not science, but myth-making.

The peppered myth?

Almost every textbook that deals with evolution not only re-tells the classical peppered moth story without mentioning its flaws, but also illustrates it with staged photographs. For example, the 2000 edition of Kenneth Miller and Joseph Levine's *Biology* includes faked photographs of peppered moths on tree trunks, and calls Kettlewell's work "a classic demonstration of natural selection in action." Similarly, Burton Guttman's 1999 *Biology* includes the usual photos, summarizes Kettlewell's experiments, and calls the peppered moth "a classic contemporary case of natural selection."

Many textbooks repeat the myth that the presence or absence of lichens was a key factor in the story. In his 1998 textbook, *Biology: Visualizing Life*, George Johnson wrote: "Recently, England has introduced strict air-pollution control measures. Forests near industrial centers like Birmingham are once again becoming covered with lichens. Have students predict what Kettlewell would find today." The 1998 edition of Cecie Starr and Ralph Taggart's *Biology: The Unity and Diversity of Life* includes the following: "In 1952, strict pollution controls went into effect. Lichens made comebacks. Tree trunks became free of soot, for the most part. As you might have predicted, directional selection started to operate in the reverse direction."

A Canadian textbook-writer who knew that peppered moth pictures were staged used them anyway. "You have to look at the audience. How convoluted do you want to make it for a first time learner?" Bob Ritter was quoted as saying in the April 5, 1999, *Alberta Report Newsmagazine*. High school students "are still very concrete in the way they learn," continued Ritter. "The advantage of this example of natural selection is that it is extremely visual." (Visual perhaps, but untrue.) Ritter explained: "We want to get across the idea of selective adaptation. Later on, they can look at the work critically."

Apparently, the "later on" can be *much* later. When University of Chicago Professor Jerry Coyne learned of the flaws in the classical story in 1998, he was well into his career as an evolutionary biologist. His experience illustrates how insidious the icons of evolution really are, since they mislead even professionals. Coyne was understandably "embarrassed" when he finally learned that the peppered moth story he had been teaching for years was a myth.

Coyne's reaction upon learning the truth reveals the disillusionment that may become increasingly common as biologists discover that the icons of evolution misrepresent the truth. "My own reaction," he wrote, "resembles the dismay attending my discovery, at the age of six, that it was my father and not Santa who brought the presents on Christmas Eve."

CHAPTER 8

Darwin's Finches

A quarter of a century before Darwin published *The Origin of Species*, he was formulating his ideas about living things as a naturalist aboard the British survey ship *H.M.S. Beagle*. The *Beagle* left England in 1831 on a five-year voyage to chart the waters of South America, and in 1835 it visited the Galápagos Islands in the Pacific, about six hundred miles off the west coast of Ecuador.

While the *Beagle* was in the Galápagos, Darwin collected specimens of the local wildlife, including some finches. Thirteen species of finches are scattered among the two dozen or so volcanic islands. (A fourteenth species lives on Cocos Island, almost four hundred miles northeast of the Galápagos.) The finches differ mainly in the size and shape of their beaks, and it is thought that they descended from birds that arrived from the mainland in the distant past.

In Darwin's theory, a single species diverges into several varieties, then into several different species, through the action of natural selection. Since the beaks of the Galápagos finches are adapted to the different foods they eat, it seems reasonable to suppose that the various species are a result of natural selection. In fact, they seem like such a good example of Darwinian evolution that they are now known as "Darwin's finches." (Figure 8-1) Many biology textbooks explain that the Galápagos finches were instrumental in helping Darwin to formulate his theory of evolution, and that field observations in the 1970s provided evidence for the theory by showing how natural selection affects the birds' beaks.

Yet the Galápagos finches had almost nothing to do with the formulation of Darwin's theory. They are not discussed in his diary of the *Beagle* voyage except for one passing reference, and they are never mentioned in *The Origin of Species*. The natural selection observed in the 1970s reversed direction in the following years, so there was no net evolutionary change. And several finch species may now be merging through hybridization—the opposite of what one would expect from the branching-tree pattern of Darwinian evolution.

The legend of Darwin's finches

While Darwin was in the Galápagos Islands, he collected nine of the thirteen species that now bear his name, but he identified only six of them as finches. Except in two cases, he failed to observe any differences in their diets, and even in those cases he failed to correlate diet with beak shape. In fact, Darwin was so unimpressed by the finches that he made no effort while in the Galápagos to separate them by island. Only after the *Beagle* returned to England did ornithologist John Gould begin to sort out their geographical relationships, and much of the information Darwin provided turned out to be wrong. Eight of the fifteen localities he recorded are in serious doubt, and most had to be reconstructed from the more carefully labeled collections of his shipmates.

Thus, according to historian of science Frank Sulloway, Darwin "possessed only a limited and largely erroneous conception of both the feeding habits and the geographical distribution of these birds." And as for the claim that the Galápagos finches impressed Darwin as evidence of evolution, Sulloway wrote, "nothing could be further from the truth."

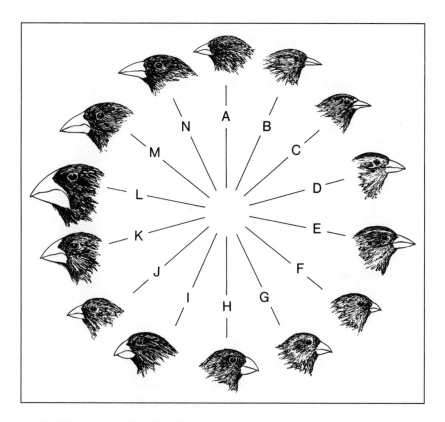

FIGURE 8-1 Darwin's finches.

The fourteen species of Darwin's finches. All live on the Galápagos Islands except (B), the Cocos Island finch. The medium ground finch (K) is the species that has been most intensively studied. Note the differences in their beaks.

In fact, Darwin did not become an evolutionist until many months after his return to England. Only years later did he look back at the finches and reinterpret them in the light of his new theory. In 1845 he wrote in the second edition of his *Journal of Researches:* "The most curious fact is the perfect gradation in the size of the beaks of the different species of [finches]. Seeing this gradation and diversity of structure in one small, intimately related group of birds, one might really fancy that from an original paucity of birds in this archipelago, one species had been taken and modified for different ends." But this was a speculative afterthought, not an inference from evidence he collected. Indeed, the confusion surrounding the geographical labeling of Darwin's specimens made it impossible for him to use them as evidence for his theory.

Nor did Darwin have a clear idea of the finches' ancestry. We now know that the thirteen species resemble each other more than they resemble any birds in Central or South America, suggesting that they may be descendants of a common ancestor that colonized the islands in the distant past. But Darwin did not visit the western coast of South America north of Lima, Peru, so for all he knew the finches were identical to species still living on the mainland.

It wasn't until the rise of neo–Darwinism in the 1930s that the Galápagos finches were elevated to their current prominence. Although they were first called "Darwin's finches" by Percy Lowe in 1936, it was ornithologist David Lack who popularized the name a decade later. Lack's 1947 book, *Darwin's Finches,* summarized the evidence correlating variations in finch beaks with different food sources, and argued that the beaks were adaptations caused by natural selection. In other words, it was Lack more than Darwin who imputed evolutionary significance to the

Galápagos finches. Ironically, it was also Lack who did more than anyone else to popularize the myth that the finches had been instrumental in shaping Darwin's thinking.

Darwin's finches as an icon of evolution

When Lack elevated the Galápagos finches to iconic status, Darwin's meager contribution to our knowledge of them grew with each re-telling of the story. According to Sulloway, "Darwin was increasingly given credit after 1947 for finches he never saw and for observations and insights about them he never made." In the most extreme form of the legend, Darwin is said to have "collected species and observed behavioral traits, such as the remarkable tool-using habit of the woodpecker finch, that were not even known in his lifetime." Thus iconography becomes hagiography.

Although Sulloway exploded the legend almost twenty years ago, many modern biology textbooks still claim that the Galápagos finches inspired Darwin with the idea of evolution. Gould and Keeton's *Biological Science* (1996) informs students that the finches "played a major role in leading Darwin to formulate his theory of evolution by natural selection." According to Raven and Johnson's *Biology* (1999), "the correspondence between the beaks of the 13 finch species and their food source immediately suggested to Darwin that evolution had shaped them." And George Johnson's *Biology: Visualizing Life* (1998) maintains that "Darwin attributed the differences in bill size and feeding habits among these finches to evolution that occurred after their ancestor migrated to the Galápagos Islands." Johnson's textbook even tells students to "imagine themselves in Darwin's place" and "write journal pages that Darwin could have written."

Yet as far as Charles Darwin's contribution is concerned, the "Darwin" in Darwin's finches is largely mythical. It wasn't until almost a century after Darwin that they assumed their present status as icons of evolution. Of course, if they really were good evidence for Darwin's theory, they might deserve their iconic status anyway.

Evidence for evolution?

If Darwin's theory is correct, then the ancestral finches that colonized the Galápagos in the distant past presumably scattered to the various islands, where they were exposed to different environmental conditions. Birds on different islands probably encountered differences in food supply, leading to natural selection on their eating apparatus—their beaks. Theoretically, this process could have led over time to the beak differences that now characterize thirteen separate species.

This is a plausible scenario, but the evidence that Lack cited for it was indirect. Differences in finch beaks are correlated with different food sources, and the birds are scattered among the various islands (though it is not the case that each island has its own species). The pattern seems to fit Darwin's theory, yet the case would be much stronger if there were some direct evidence for the process.

One sort of direct evidence could be genetic. But apart from knowing that finch beaks are highly heritable—that the beak of a finch is very likely to resemble the beaks of its biological parents— we know nothing about the genetics of finch beaks. Chromosome studies show no differences among the Galápagos finches, and the DNA studies that have been used to construct molecular phylogenies relied on genes unrelated to beak shape.

Another sort of direct evidence would be observations of natural selection in the wild. This evidence has been supplied by the husband-and-wife team of Peter and Rosemary Grant, who went to the Galápagos in the 1970s to observe evolution in action.

The beak of the finch

The Grants made their first trip to the Galápagos in 1973. With the help of several other biologists, the Grants set about catching and banding finches on seven of the islands. Each finch was carefully measured for body weight, the lengths of its wings, legs and toes, and the length, width, and depth of its beak. There was variation among the finches in all these features—especially the beaks.

By 1975 the Grants and their colleagues had focused their attention on one of the smaller islands, Daphne Major. (Figure 8-2) Its small size made Daphne Major an ideal natural laboratory where they were able to band and measure every individual in one particular species, the medium ground finch. (Figure 8-1, K) The biologists even recorded matings, and banded and observed the offspring. They also kept track of rainfall, and how many seeds were produced by the island's plant species.

During the early 1970s Daphne Major received regular rainfall that supported an abundant food supply and a large finch population. In normal rainy seasons, such as that of 1976, the island received about 5 inches of rain; but in 1977 only about an inch fell. The 1977 drought caused a severe reduction in the availability of seeds, and the island's population of medium ground finches declined to about 15 percent of its former size. The Grants and their colleagues observed that survivors of the drought tended to have slightly larger bodies and slightly larger beaks. They also noted that the supply of small seeds was drastically reduced that year. They concluded that natural selection had strongly favored those birds capable of cracking the tough, large seeds that remained.

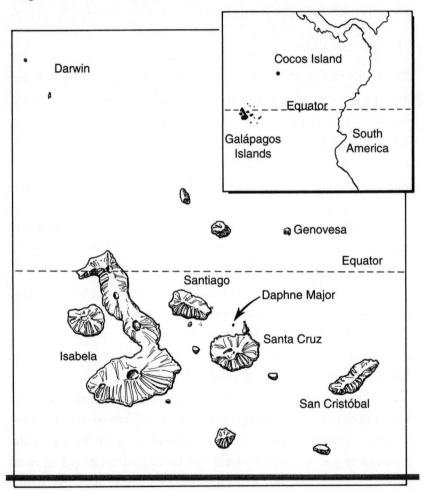

FIGURE 8-2   The Galápagos Islands.

The Grants' pioneering work on finch beaks took place mostly on Daphne Major, a tiny island just north of Santa Cruz.

As a result of the drought, the average beak depth of medium ground finches increased about 5 percent. (Beak depth is the distance between the top and bottom of the beak at its base.) This amounted to a difference of about half a millimeter—the thickness of a human thumbnail. This may not seem like much, but for the finches on Daphne Major in 1977 it meant the difference between life and death.

It was also a dramatic example of natural selection in the wild. The story of the Grants' research was recounted in Jonathan Weiner's 1994 book, *The Beak of the Finch*, which called the observed change in beak depth "the best and most detailed demonstration to date of the power of Darwin's process." Because of this, according to Weiner, the beak of the finch is "an icon of evolution."

The Grants and their colleagues realized at the time that natural selection might oscillate between dry and wet years, making beaks larger one year and smaller the next. But if beak depth were to continue increasing, then something very interesting might happen. The various species of Darwin's finches are distinguished mainly by differences in their beaks. The Grants reasoned that if natural selection can produce changes in beaks, perhaps it could also explain the origin of species among Darwin's finches.

In *Scientific American* in 1991, Peter Grant explained how this could happen, at least in theory. Calling the increase in beak depth during severe drought a "selection event," Grant estimated the number of such events required to transform the medium ground finch into another species: "The number is surprisingly small: about 20 selection events would have sufficed. If droughts occur once a decade, on average, repeated directional selection at this rate with no selection in between droughts would transform one species into another within 200 years. Even if the estimate is off by a factor of 10, the 2,000 years required for speciation is still very little time in relation to the hundreds of thousands of years the finches have been in the archipelago."

Grant's extrapolation depends, of course, on the assumption that increases in beak size are cumulative from one drought to the next. But the Grants and their colleagues knew that this is not the case.

When the rains returned

People who live on the west coast of North or South America know that every few years they can expect an El Niño—a disturbance in winter weather patterns caused by unusually warm air over the Pacific Ocean. In the winter of 1982-1983, an El Niño brought heavy rains to the Galápagos Islands—over ten times more than normal, and fifty times more than fell during the drought. Plant life exploded, and so did the finch population.

After the 1982-1983 El Niño, with food once again plentiful, the average beak size in medium ground finches returned to its previous value. In 1987 Peter Grant and his graduate student, Lisle Gibbs, reported in *Nature* that they had observed "a reversal in the direction of selection" due to the change in climate. "Large adult size is favoured when food is scarce," they wrote, "because the supply of small and soft seeds is depleted first, and only those birds with large bills can crack open the remaining large and hard seeds. In contrast, small adult size is favoured in years following very wet conditions, possibly because the food supply is dominated by small soft seeds."

So the evolutionary change that the Grants and their colleagues had observed during the drought of 1977 was reversed by the heavy rains of 1983. "Selection had flipped," wrote Weiner. "The birds took a giant step backward, after their giant step forward." As Peter Grant wrote in 1991, "the population, subjected to natural selection, is oscillating back and forth" with every shift in climate.

By itself, however, oscillating selection cannot produce any net change in Darwin's finches, no matter how long it continues.

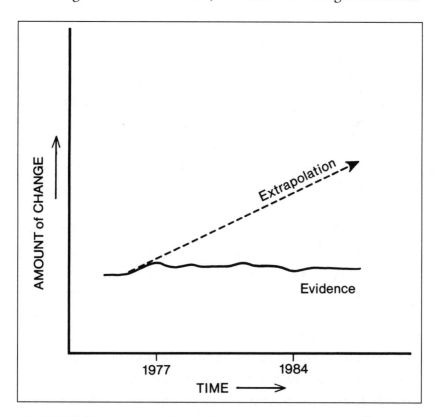

FIGURE 8-3   A comparison of straight-line versus cyclical change.

The straight line represents the extrapolation that predicts the origin of a new species of finch in two hundred years. The wavy line represents the cyclical changes so far observed.

(Figure 8-3) Some sort of long-term trend would have to be superimposed on the back-and-forth oscillations to produce long-term change, and that is not what the Grants and their colleagues witnessed. Indeed, it would probably take much longer than a decade or two to measure it, even if it were present. Of course, the climate of the Galápagos might change in the future and alter the pattern. But both of these—an unseen trend and a future change—are speculations.

It remains a theoretical possibility that the various species of Galápagos finches originated through natural selection. But the Grants' observations provided no direct evidence for this. And in the course of their work, they discovered that several species of Darwin's finches may now be merging rather than diverging.

Diverging or merging?

If Darwinian evolution requires that one population diverge into two, the opposite would be for two previously separate populations to merge into one. (Figure 8-4) Yet this may now be happening to several species of Darwin's finches.

At least half of the finch species on the Galápagos are known to hybridize, though they do so infrequently. In the years following the 1982-1983 El Niño, the Grants and their colleagues noticed that several finch species on one island were producing hybrids that not only thrived, but also reproduced successfully. In fact, the hybrids did better than the parental species that produced them. The Grants noted that this process, if unchecked, "should lead to fusion of the species into one population." This would not happen overnight: Extrapolating from the observed frequency of hybridization, the Grants estimated that it would take one hundred to two hundred years for these species to merge completely.

So if we extrapolate from processes observed in the present, we obtain two contradictory predictions: unchecked selection for larger beaks could produce speciation in two hundred to two thousand years, while unchecked hybridization could produce the opposite of speciation in one hundred to two hundred years. Clearly, the tendency to diverge is more than offset by the tendency to merge. Of course, the fluctuating climate of the Galápagos means that neither process is likely to continue indefinitely, and the Grants concluded that "over the long term there should be a selection-hybridization balance." According to Weiner it seems that a "vast, invisible pendulum [is] swinging back and forth in Darwin's islands, an oscillation with two phases," in which the finches "are perpetually being forced slightly apart and drifting back together again."

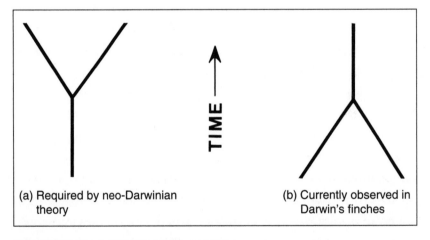

FIGURE 8-4 **Diverging vs. merging.**

(a) The splitting of one species into two, as required for Darwinian evolution. (b) The merging of two species together due to hybridization, currently being observed in several species of Darwin's finches.

So Darwin's finches may not be merging or diverging, but merely oscillating back and forth. Their success at hybridizing, however, raises a question about whether they are separate species at all.

Fourteen species, or six?

It turns out that most of the fourteen species of Darwin's finches—or at least most of the thirteen living on the Galápagos Islands—remain distinct primarily because of mating behavior. Evidence suggests that the birds choose their mates on the basis of beak morphology and song pattern. The former is inherited, while the latter is learned by young birds from their parents.

But one might expect that true species would be separated by more than beak morphology and song pattern. In human populations, race is inherited and language is learned—just as, in finches, beaks are inherited and songs are learned. Yet human populations that are separated by race and language are unquestionably part of the same species, even though such differences may make interbreeding uncommon.

Writing in *Science* in 1992, the Grants noted that the superior fitness of hybrids among populations of Darwin's finches "calls into question their designation as species." The following year, Peter Grant acknowledged that if species were strictly defined by inability to interbreed then "we would recognize only two species of Darwin's finch on Daphne," instead of the usual four. "The three populations of ground finches on Genovesa would similarly be reduced to one species," Grant continued. "At the extreme, six species would be recognized in place of the current 14, and additional study might necessitate yet further reduction."

In other words, Darwin's finches may not be fourteen separate species. Perhaps they are in the process of *becoming* species.

But then we would expect their tendency to diverge through natural selection to be greater than their tendency to merge through hybridization, and this is not what the evidence shows. Perhaps the Galápagos finches *used* to be separate species and are now in the process of becoming fewer. But then they demonstrate the *opposite* of Darwinian evolution, which occurs when one species divides into separate species.

The increase in average beak size in several species of Galápagos finches after a severe drought—and its return to normal after the drought ended—is direct evidence for natural selection in the wild. In this limited sense, the finches provide evidence for Darwin's theory. As examples of the origin of species by natural selection, however, Darwin's finches leave a lot to be desired—though this hasn't stopped some people from using them as examples anyway. But the only way they can do this is by exaggerating the evidence.

Exaggerating the evidence

Thanks to years of careful research by the Grants and their colleagues, we know quite a lot about natural selection and breeding patterns in Darwin's finches. And the available evidence is clear. First, selection oscillates with climatic fluctuations, and does not exhibit long-term evolutionary change. Second, the superior fitness of hybrids means that several species of Galápagos finches might be in the process of merging rather than diverging.

The Grants' excellent field work provided us with a good demonstration of natural selection in the wild—far better than Kettlewell's peppered moths. If the Grants had stopped there, their work might stand as an example of science at its best. Yet they have tried to make more of their work than the evidence warrants. In articles published in 1996 and 1998, the Grants declared that the Darwinian theory of the origin of species "fits the facts of Darwin's Finch evolution on the Galápagos Islands," and that "the driving force" is natural selection.

This claim was echoed by Mark Ridley in his 1996 college textbook, *Evolution*. Like the Grants, Ridley extrapolated the increase in beak size after the 1977 drought to estimate the time it would take to produce a new species. This "illustrates how we can extrapolate from natural selection operating within a species to explain the diversification of the finches from a single common ancestor." Ridley concluded: "Arguments of this kind are common in the theory of evolution."

Indeed. But arguments of this kind exaggerate the truth. And this exaggeration seems to characterize many claims for Darwin's theory. Evidence for change in peppered moths is claimed as evidence for natural selection even though the selective agent has not been demonstrated. And evidence for oscillating natural selection in finch beaks is claimed as evidence for the origin of finches in the first place. Apparently, some Darwinists are prone to make inflated claims for rather meager evidence.

Does the National Academy of Sciences endorse "arguments of this kind" that exaggerate the evidence? A 1999 booklet published by the National Academy describes Darwin's finches as "a particularly compelling example" of the origin of species. The booklet goes on to explain how the Grants and their colleagues showed "that a single year of drought on the islands can drive evolutionary changes in the finches," and that "if droughts occur about once every 10 years on the islands, a new species of finch might arise in only about 200 years."

That's it. Rather than confuse the reader by mentioning that selection was reversed after the drought, producing no long-term evolutionary change, the booklet simply omits this awkward fact. Like a stock promoter who claims a stock might double in value in twenty years because it increased 5 percent in 1998, but doesn't mention that it decreased 5 percent in 1999, the booklet misleads the public by concealing a crucial part of the evidence.

This is not truth-seeking. It makes one wonder how much evidence there really is for Darwin's theory. As Berkeley law professor and Darwin critic Phillip E. Johnson wrote in *The Wall Street Journal* in 1999: "When our leading scientists have to resort to the sort of distortion that would land a stock promoter in jail, you know they are in trouble."

CHAPTER 9

Four-Winged
Fruit Flies

I n Darwin's theory, evolution is a product of two factors: nat-ural selection and heritable variation. Natural selection molds populations by preserving favorable variations that are passed on to succeeding generations. Small-scale evolution within a species (such as we see in domestic breeding) makes use of variations already present in a population, but large-scale evolution (such as Darwin envisioned) is impossible unless new variations arise from time to time. Darwin devoted the first two chapters of *The Origin of Species* to establishing the existence of heritable variations in domestic and wild populations, but he did not know how they are inherited or how new ones arise.

It wasn't until the advent of neo-Darwinism and molecular genetics in the twentieth century that many biologists finally felt they understood the mechanism of heredity and the origin of variations. According to modern neo-Darwinism, genes con-sisting of DNA are the carriers of hereditary information; infor-mation encoded in DNA sequences directs the development of the organism; and new variations originate as mutations, or acci-dental changes in the DNA.

Some DNA mutations have no effect, and most others are harmful. Occasionally, however, a mutation comes along that is beneficial—it confers some advantage on an organism, which can then leave more offspring. According to neo-Darwinism, beneficial DNA mutations—though not needed for limited modifications within a species—provide the raw materials necessary for large-scale evolution.

Beneficial mutations are rare, but they do occur. For example, mutations can have biochemical effects that render bacteria resistant to antibiotics or insects resistant to insecticides. But biochemical mutations cannot explain the large-scale changes in organisms that we see in the history of life. Unless a mutation affects morphology—the shape of an organism—it cannot provide raw materials for morphological evolution.

One organism in which morphological mutations have been extensively studied is the fruit fly, *Drosophila melanogaster*. Among the many mutations that are now known in *Drosophila*, some cause the normally two-winged fruit fly to develop a second pair of wings. Since 1978, the four-winged fruit fly has become increasingly popular in textbooks and public presentations as an icon of evolution. (Figure 9-1)

But four-winged fruit flies do not occur spontaneously. They must be carefully bred in the laboratory from three artificially maintained mutant strains. Furthermore, the extra wings lack flight muscles, so the mutant fly is seriously handicapped. Four-winged fruit flies testify to the skill of geneticists, and they help us to understand the role of genes in development, but they provide no evidence that DNA mutations supply the raw materials for morphological evolution.

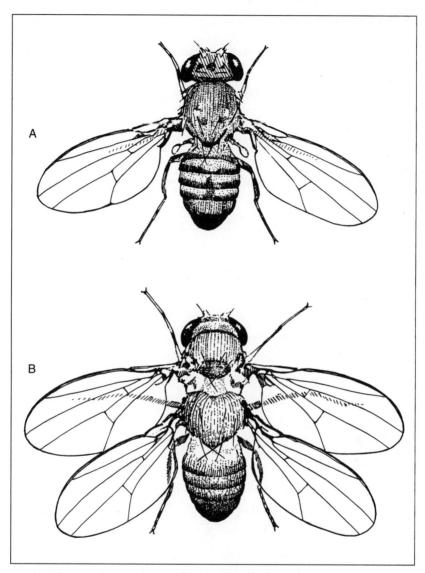

FIGURE 9-1   Normal and four-winged fruit flies.

(A) A normal or "wild-type" fruit fly, with two wings and two balancers or "halteres" (tiny appendages on either side between the wings and the rear legs). (B) A mutant fly in which the halteres have developed into normal-looking wings.

The origin of variations from Darwin to DNA

Although Darwin did not know the origin of variations, he believed that "changed conditions of life are of the highest importance" in causing them. In other words, he thought that most new variations are induced by the environment, acting either on the whole organism or on its reproductive system. In some cases, he wrote, new heritable variations "may be attributed to the increased use or disuse of parts."

This view, known as the inheritance of acquired characteristics, had been advocated a half century earlier by the French zoologist Jean Baptiste de Lamarck. It wasn't until the last years of Darwin's life (he died in 1882) that German zoologist August Weismann persuaded most biologists that Lamarck's view was false. According to Weismann, inherited characteristics are transmitted by "germ cells" that remain separate from the rest of the body from the embryo through adulthood, when they give rise to eggs or sperm. In a famous experiment, he cut off the tails of several generations of mice to prove that disuse did not produce mice with shorter tails.

The biological basis of heredity remained unknown, however, until Gregor Mendel's theory became generally known after 1900. Cell biologists identified chromosomes as the carriers of Mendel's heredity factors, and in 1909 Wilhelm Johanssen named them "genes." In the days before DNA, genes were regions on chromosomes, and American fruit fly geneticist Thomas Hunt Morgan studied spontaneous changes in individual genes that he called mutations (a term he borrowed from Dutch botanist Hugo DeVries).

By the 1930s many geneticists believed that the sort of mutations Morgan studied were the source of new variations needed for evolution. In 1937 Theodosius Dobzhansky made this a fundamental tenet of neo-Darwinism when he wrote that "mutations and chromosomal changes…constantly and unremittingly supply the raw materials for evolution." In the 1940s microbiologists showed that DNA carries hereditary information, and in 1953 James Watson and Francis Crick explained how the molecular structure of DNA might determine and transmit heritable traits. Morgan's mutations were attributed to molecular accidents, and the picture seemed complete. In 1970, molecular biologist Jacques Monod announced that "the mechanism of Darwinism is at last securely founded."

We now know that some DNA mutations are "neutral"—they have no effect at all. The vast majority of the rest are harmful. In the struggle for existence, natural selection would be expected to ignore the former and eliminate the latter. Only those rare mutations which benefit the organism could be favored by natural selection, and thus provide raw materials for evolution. Some mutations that affect biochemical pathways fit this description.

Beneficial biochemical mutations

Antibiotics work by poisoning molecules in bacteria. Most cases of medically significant antibiotic resistance are not due to mutations, but to complex enzymes that inactivate the poison, and which bacteria either inherit or acquire from other organisms. Some cases of resistance, however, are due to spontaneous mutations that alter the bacteria's molecules just enough so an antibiotic can no longer poison them. Bacteria lucky enough to have such mutations (like those lucky enough to have inactivating enzymes) can resist an antibiotic and survive to reproduce.

Like antibiotic resistance, most insecticide resistance is due to inactivating enzymes. There are cases, however, in which resistance is due to spontaneous mutations. Like the mutations that confer resistance to antibiotics, these can benefit the organism by enabling it to survive and reproduce despite the presence of the poison.

Since mutations leading to antibiotic and insecticide resistance are clearly beneficial in certain environments, biology textbooks invariably list them as evidence that mutations provide the raw materials for evolution. Many textbooks also list sickle-cell anemia, because the same mutation that causes this crippling genetic disease can also, in a milder form, benefit infants growing up in malaria-ridden areas. In all of these cases, however, the evolution that occurs is trivial. The raw materials for large-scale evolution must be able to contribute to fundamental changes in an organism's shape and structure.

Since biochemical mutations—such as those leading to antibiotic resistance and sickle-cell anemia—do not affect an organism's shape or structure, evolution needs beneficial mutations that affect morphology. Neo-Darwinists know this, of course, and to provide evidence of morphological mutations a growing number of them are using pictures of mutant fruit flies with an extra pair of wings.

The four-winged fruit fly

The bodies of fruit flies consist of segments, three of which are in the thorax (midsection). Normally, the second thoracic segment bears a pair of wings, and the third bears a pair of "halteres," or balancers—tiny appendages that enable the insect to maintain its balance in flight. (Figure 9–1a) In 1915 geneticist

Calvin Bridges (working in Thomas Hunt Morgan's laboratory) discovered a mutant fruit fly in which the third thoracic segment looked a bit like the second, and the halteres were slightly enlarged and looked like miniature winglets. This spontaneously occurring "*bithorax*" mutant has been maintained as a laboratory stock ever since.

In 1978 California Institute of Technology geneticist Ed Lewis reported that by breeding flies possessing the *bithorax* mutation with flies possessing another mutation, "*postbithorax*," he was able to produce a fruit fly in which the halteres were even more enlarged, and looked almost like a second pair of wings. He subsequently found that if flies combining these two mutations were bred with flies possessing a third, "*anterobithorax*," the triple-mutant offspring had an extra pair of wings that looked like the fly's normal wings. (Figure 9-1, B)

Lewis had to use three mutations because no single mutation affected the entire segment. Each fruit fly segment is divided into an anterior (forward) compartment and a posterior (rearward) compartment. The *postbithorax* mutation induced the posterior compartment of the third thoracic segment to produce the rear half of a wing, while the combination of *anterobithorax* and *bithorax* mutations caused the anterior compartment to produce the forward half of a wing. Only a fly possessing all three mutations bears four normal-looking wings. (Figure 9-2)

Of course, Lewis's goal was not to produce sideshow freaks, but to understand the molecular interactions involved in fruit fly development. It turns out that all three mutations in the four-winged fruit fly affect a single large gene, "*Ultrabithorax*." The mutations do not affect the protein produced by the gene, but only where the protein is produced. Every cell in the fruit fly's body receives the same genes from the fertilized egg; but as the

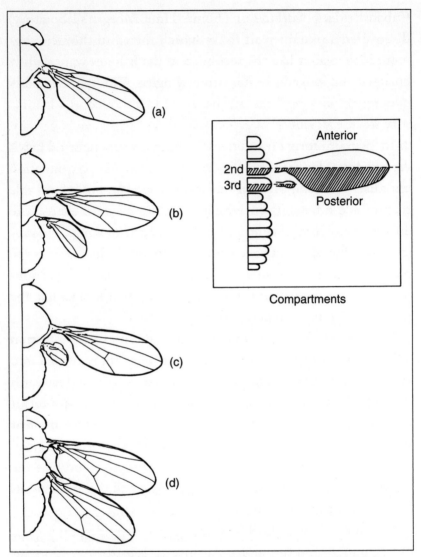

FIGURE 9-2   Steps in the construction of a four-winged fruit fly.

The box at the upper right shows how each segment is divided into an anterior and posterior compartment. (a) Normal fly; (b) *bithorax* mutant; (c) *postbithorax* mutant; (d) triple mutant (*anterobithorax, bithorax,* and *postbithorax*). The *anterobithorax* mutation enhances the effect of *bithorax*.

embryo develops, specific genes are turned on only in those cells where they are needed. This process depends on "regulatory sequences" associated with each gene. Such sequences act like switches, allowing genes to be turned on or off in different parts of the embryo.

In a normal fruit fly, the *Ultrabithorax* gene is turned *on* in the third thoracic segment, and the segment produces halteres rather than wings. The *anterobithorax, bithorax,* and *postbithorax* mutations each turn the gene *off* to some degree: The first two turn it off in the anterior compartment, and the third turns it off in the posterior compartment. When all three mutations are present, the gene is completely turned off in the third thoracic segment, which then produces a pair of normal-looking wings instead of halteres.

By deciphering the genetic interactions involved in turning off *Ultrabithorax,* Lewis was able to shed considerable light on the molecular biology of fruit fly development, and his research earned him a Nobel Prize in 1995. But how much light do four-winged fruit flies shed on evolution?

Four-winged fruit flies and evolution

According to Peter Raven and George Johnson's 1999 textbook, *Biology,* "all evolution begins with alterations in the genetic message... Genetic change through mutation and recombination [the re-arrangement of existing genes] provides the raw materials for evolution." The same page features a photo of a four-winged fruit fly, which is described as "a mutant because of changes in *Ultrabithorax,* a gene regulating a critical stage of development; it possesses two thoracic segments and thus two sets of wings."

The textbook does not explicitly claim that the four-winged fruit fly shows us evolution in action, but it uses the fly in its discussion of evolution to imply that genetic mutations are the origin of new variations. The textbook fails to explain, however, that three separate mutations had to be artificially combined in one fly to produce a second set of normal-looking wings. Such a combination is exceedingly unlikely to occur in nature.

Even more seriously, the textbook fails to point out that the second pair of wings is non-functional. Biologists have known since the 1950s that the extra wings on *bithorax* mutants lack flight muscles. The hapless insect is thus disabled, and the disability increases with the size of the mutant appendages. In aerodynamic terms, a triple-mutant four-winged fruit fly is like an airplane with an extra pair of full-sized wings dangling loosely from its fuselage. It may be able to get off the ground, but its flying ability is seriously impaired. Because of this, four-winged males have difficulty mating, and unless the line is carefully maintained in a laboratory it quickly dies out.

So four-winged fruit flies are not raw materials for evolution. Even neo-Darwinists acknowledge this. Ernst Mayr wrote in 1963 that major mutations such as *bithorax* "are such evident freaks that these monsters can be designated only as 'hopeless.' They are so utterly unbalanced that they would not have the slightest chance of escaping elimination" through natural selection. In addition, finding a suitable mate for the "hopeless monster" seemed to Mayr to be an insurmountable difficulty. Given this long-standing objection to the evolutionary significance of such monsters, the recent popularity of four-winged fruit flies is puzzling. Perhaps, like pictures of peppered moths on tree trunks, they are just too "visual" to resist.

Adding to the confusion, textbook accounts typically leave the reader with the impression that the extra wings represent a *gain* of structures. But four-winged fruit flies have actually *lost* structures which they need for flying. Their balancers are gone, and instead of being replaced with something new have been replaced with copies of structures already present in another segment. Although pictures of four-winged fruit flies give the impression that mutations have added something new, the exact opposite is closer to the truth.

Someone attempting to salvage these mutants as evidence for neo-Darwinism might point out that even a loss of structures can have evolutionary significance. And indeed it can. Evolutionary biologists believe that two-winged flies evolved from four-winged flies. It is conceivable that ancestral four-winged flies acquired genetic mutations which reduced one pair of wings to tiny rudiments, and these became halteres. Perhaps *bithorax* is showing us mutations back to the ancestral state—in other words, evolution in reverse. This scenario is plausible, but once again the evidence points in the wrong direction.

Evolution in reverse?

In support of the view that two-winged flies evolved from four-winged flies, a 1998 booklet published by the National Academy of Sciences points out that "geneticists have found that the number of wings in flies can be changed through mutations in a single gene." Although this statement is technically true, it is quite misleading—and not just because three separate mutations are necessary and the extra wings are nonfunctional.

What really changes the number of wings in a fly is a complex genetic network. A four-winged fly does not become a two-winged fly because mutations knock out some hypothetical "wing gene," but because the fly acquires a whole network of developmental controls that transform one set of wings into functional halteres.

The *Ultrabithorax* gene itself is large and complex. It consists of about a hundred thousand DNA subunits, most of which are involved in regulating when and where the gene is turned on in the embryo. And *Ultrabithorax* does not function alone. In 1998 Scott Weatherbee and a team of developmental biologists reported that *Ultrabithorax* affects haltere development "by independently regulating selected genes that act at different levels of the wing patterning hierarchy." It is this entire hierarchy, and not just one gene, that had to evolve in order to convert wings into halteres. According to Weatherbee and his colleagues, "the evolution of the haltere progressed through the accumulation of a complex network of [*Ultrabithorax*]-regulated interactions." Biologists do not understand how fruit flies acquired this complex network, but it certainly could not have originated from just a few mutations in a single gene.

What the four-winged fruit fly shows us is that mutations can shut down a complex network of interactions. But there's nothing surprising about this; we know that a single mutation can shut down an entire embryo and kill it outright. Damaging a complex regulatory network with mutations doesn't explain how the network originated, any more than killing an embryo with a lethal mutation explains how flies evolved. Yet it is precisely the *origin* of the network that we need to understand if we are to explain how four-winged flies evolved into two-winged flies.

So the four-winged fruit fly is a useful window on the genetics of development, but it provides no evidence that mutations supply the raw materials for morphological evolution. It does not even show us evolution in reverse. As evidence for evolution, the four-winged fruit fly is no better than a two-headed calf in a circus sideshow.

Why, then, has it become popular to feature the four-winged fruit fly in textbooks and public presentations defending Darwin's theory? Could it be concealing a deeper problem with the evidence for neo-Darwinism?

Are DNA mutations the raw materials for evolution?

According to biology textbooks, DNA mutations are unquestionably the source of new variations for evolution. For example, the 1998 edition of Cecie Starr and Ralph Taggart's *Biology: The Unity and Diversity of Life* tells students that "every so often, a new mutation bestows an advantage on the individual... beneficial mutations, and neutral ones, have been accumulating in different lineages for billions of years. Through all that time, they have been the raw material for evolutionary change—the basis for the staggering range of biological diversity, past and present." Burton Guttman's 1999 textbook, *Biology*, declares that "*mutation is ultimately the source of all genetic variation and therefore the foundation for evolution.*" (emphasis in original)

Yet the evidence cited in these textbooks falls far short of supporting these sweeping claims. To be sure, biochemical mutations lead to antibiotic and insecticide resistance, and human beings carrying the sickle-cell trait are more likely to survive malaria as infants. But only beneficial *morphological* mutations can provide raw materials for morphological evolution, and evidence for such mutations is surprisingly thin. As we have seen, four-winged fruit flies do not provide the missing evidence, despite their current popularity.

If textbook-writers have no good examples of beneficial morphological mutations, it's not because biologists haven't been looking for them. About the time that Lewis was studying *Ultrabithorax*, German geneticists Christiane Nüsslein-Volhard and Eric Wieschaus were using a technique called "saturation mutagenesis" to search for every possible mutation involved in fruit fly development. They discovered dozens of mutations that affect development at various stages and produce a variety of malformations. Their Herculean efforts earned them a Nobel prize (which they shared with Lewis), but they did not turn up a single morphological mutation that would benefit a fly in the wild.

Saturation mutagenesis has also been used in a tiny worm studied by many developmental biologists, and is currently being applied to zebrafish. So far, no morphological mutations that would be beneficial in nature have been found in these animals, either.

Since direct evidence has been so hard to come by, neo-Darwinists usually cite indirect evidence. Genetic differences between two organisms are taken to indicate that their morphological differences are due to changes in genes. But without direct evidence, neo-Darwinists can only *assume* that genetic differences are the cause of morphological differences. As we saw in the chapter on homology, there are many cases in which similarities and differences in genes are *not* correlated with similarities and differences in morphology. Obviously, it is reasonable to question the neo-Darwinian claim that genetic mutations are the raw materials for large-scale evolution.

But people who question the claim are likely to encounter considerable resistance from defenders of neo-Darwinism. If they persevere in their questioning, however, they will find that they are not alone, and that the problem is bigger than they imagined. According to many biologists in the past, and many non-American biologists in the present, genes are not as important as neo-Darwinists make them out to be.

Beyond the gene

Like fruit flies, human beings begin life as a single fertilized egg cell. As the egg divides, it bequeaths a full set of genes to each of its progeny. Eventually, the fertilized egg divides into several hundred types of cells: A skin cell is different from a muscle cell, which in turn is different from a nerve cell, and so on. Yet with a few exceptions, all these cell types contain the same genes as the fertilized egg.

The presence of identical genes in cells that are radically different from each other is known as "genomic equivalence." For a neo-Darwinist, genomic equivalence is a paradox: If genes control development, and the genes in every cell are the same, why are the cells so different?

According to the standard explanation, cells differ because the genes are differentially turned on or off. Cells in one part of the embryo turn on some genes, while cells in another part turn on others. This certainly happens, as we saw in the case of *Ultrabithorax*. But it doesn't resolve the paradox, because it means that genes are being turned on or off by factors outside themselves. In other words, control rests with something beyond the genes—something "epigenetic." This does not imply that mystical forces are at work, but only that genes are being regulated by cellular factors outside the DNA.

Many biologists during the first half of the twentieth century investigated epigenetic factors in their attempts to understand embryo development, but the factors proved elusive. As the neo-Darwinian synthesis of Mendelian genetics with Darwinian evolution rose to prominence between the two World Wars, biologists studying epigenesis were increasingly marginalized. According to historian Jan Sapp, American geneticists such as Thomas Hunt Morgan took "an operational approach to their work, defining heredity and the gene in terms of the experimental operations by which they might be demonstrated." They thereby opted for "rapid production of results based on studies which could be carried out easily by established procedures."

At the same time, the neo-Darwinian synthesis of genetics and evolution was becoming increasingly popular, and neo-Darwinists welcomed the gene-centered emphasis in American research. Biologists who continued the difficult search for epigenetic factors were unable to match the flood of data being turned out by genetics labs. Furthermore, as Sapp put it, their ideas "seemed to threaten the significance of the merger of Mendelian genetics and selection theory and therefore had to be denied." The operational success and doctrinal aggressiveness of American neo-Darwinists enabled them to establish a near-monopoly over academic jobs, research funding, and scientific journals that persists to this day.

But neo-Darwinian genetics never resolved the paradox of genomic equivalence. In fact, the paradox recently deepened with the discovery that developmental genes such as *Ultrabithorax* are similar in many different animals—including flies and humans. If our developmental genes are similar to those of other animals, why don't we give birth to fruit flies instead of human beings?

The paradox of genomic equivalence has been largely ignored by gene-centered American biologists, but less so by Europeans. In March 1999 I attended a conference on "Genes and Development" in Basel, Switzerland. About fifty European biologists and philosophers of science were present, all of them critical of the neo-Darwinian doctrine that genes control embryo development.

One of the speakers began her talk with some jokes about the obligatory confessions of faith in Darwinism that are expected of speakers at scientific conferences. She went on to explain that DNA sequences do not even uniquely determine the sequence of amino acids in proteins, much less the larger features of cells or embryos. During the question-and-answer session that followed, a participant pointed out that most biologists already know this. She asked: "Then why don't they say so publicly?" The participant responded that it would "reduce their chances of getting money."

Later, at lunch, the lecturer told me about an experience she had had a few months earlier at a conference in Germany. There she had made some remarks critical of neo-Darwinian evolution, after which a prominent American biologist and textbook-writer had taken her aside. He had told her that she would be wise not to criticize neo-Darwinism if she ever found herself speaking to an American audience, because they would write her off as a creationist—even though she's not. She laughed as she told me the story; obviously, she was more amused than intimidated.

I was amused, too—but also saddened. It seems that scientists in Germany, like scientists in communist China, have more freedom to criticize Darwinism than scientists in America. Yet we are constantly told that scientists welcome critical thinking, and that America treasures freedom of speech. Except, apparently, when it comes to Darwinian evolution.

CHAPTER 10

Fossil Horses and Directed Evolution

Three years before Charles Darwin's death in 1882, Yale University paleontologist Othniel Marsh published a drawing of horse fossils to show how modern one-toed horses had evolved from a small four-toed ancestor. Marsh's drawing, which included only leg-bones and teeth, was soon supplemented by skulls, and illustrations of horse fossils quickly found their way into museum exhibits and biology textbooks as evidence for evolution.

Early versions of these illustrations showed horse evolution proceeding in a straight line from the primitive ancestor through a series of intermediates to the modern horse. (Figure 10-1) But paleontologists soon learned that horse evolution was much more complicated than this. Instead of being a linear progression from one form to another, it appeared to be a branching tree, with most of its branches ending in extinction.

Although advocates of Darwinian evolution have done almost nothing to correct the other icons of evolution, they have made a determined effort to correct this one. Since the 1950s, neo-Darwinian paleontologists have been actively campaigning to replace the old linear picture of horse evolution with the branching tree.

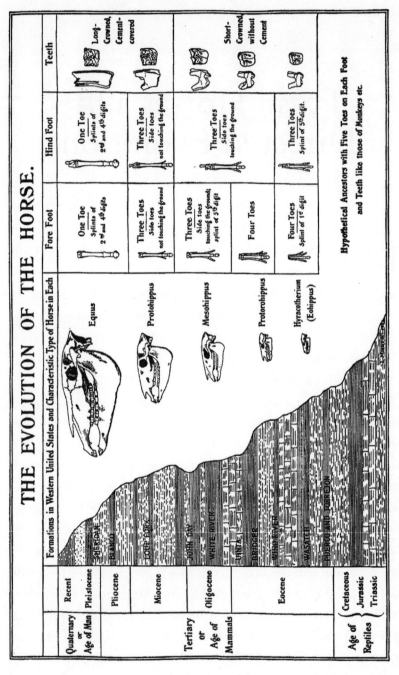

FIGURE 10-1 The old icon of horse evolution.

FIGURE 10-1   The old icon of horse evolution.

Drawings such as this one (created in 1902) used to be common in museum exhibits and biology textbooks, and can still be found in some places today. The two oldest members of the series, *Hyracotherium* and *Protorohippus*, had four toes on their front feet; the next two members, *Mesohippus* and *Protohippus*, each had three; and *Equus*, the modern horse, has one.

The reason for their campaign, however, is more interesting than the horse icon itself. People used to regard the old icon as evidence that evolution was directed, either supernaturally or by internal vital forces. Neo-Darwinists now ridicule directed evolution as a myth, and cite the new branching-tree arrangement of horse fossils as evidence that evolution is undirected.

But the doctrine of undirected evolution is philosophical, not empirical. It preceded all evidence for Darwin's theory, and it goes far beyond the evidence we now have. Like several other Darwinian claims we've seen, it is a concept masquerading as a neutral description of nature.

Fossil horses and orthogenesis

Most evolutionists who were Darwin's contemporaries believed that evolution was directed. Some regarded human beings as the divinely pre-ordained goal of the evolutionary process, while others saw evolutionary trends as directed by forces inherent in the organisms themselves. Those forces might be vital principles, or simply built-in constraints that channeled evolution in particular directions. The view that evolution was directed by internal forces or constraints became known as "orthogenesis" (from the Greek words for "straight" and "origin").

Orthogenesis was especially popular among paleontologists, because there are many trends in the fossil record that it seemed to explain. The most famous of these was the horse progression. In 1950 German paleontologist Otto Schindewolf wrote that "excellent examples of orthogenetic courses of events are provided by the progressive reduction of digits," and this process "is best and most completely known in the evolution leading to the modern horse." Schindewolf attributed orthogenesis to mechanisms inherent in the organism, rather than a supernaturally ordained goal. "It is not the conceptual final point but the concrete starting point," he explained, "that determines and brings about the orientation of evolution. Such a view can be based on actual, causative mechanisms."

But the causative mechanisms to which Schindewolf referred were never found. Meanwhile, neo-Darwinists were claiming they could explain evolution in terms of natural selection acting on random genetic mutations. Although the neo-Darwinian mechanism had not been shown to produce anything like horse evolution, it was at least clearly defined. In 1949 American paleontologist George Gaylord Simpson (one of the architects of neo-Darwinism) wrote: "Adaptation has a known mechanism: natural selection acting on the genetics of populations.... It is not quite completely understood as yet, but its reality is established and its adequacy is highly probable." Thus "we have a choice between a concrete factor with a known mechanism and the vagueness of inherent tendencies, vital urges, or cosmic goals, without known mechanism."

So orthogenesis lacked a mechanism. It also seemed to become less plausible when new evidence led to a revised picture of horse evolution.

Revising the picture of horse evolution

By the 1920s it was already becoming clear that the evolution of the horse was much more complicated than Marsh's linear picture implied. Paleontologist William Matthew and his graduate student, Ruben Stirton, established that several extinct horse species coexisted with the *Protohippus,* and that the history of horses ranged back and forth over several continents. The fossil record of horses looked less like a straight line and more like Darwin's branching tree. (Figure 10-2)

In 1944 Simpson wrote that the "general picture of horse evolution is very different from most current ideas of orthogenesis." In particular, its branching-tree pattern is "flatly inconsistent with the idea of any inherent rectilinearity." Furthermore, the trends that had seemed to support orthogenesis were illusory. For example, the trend toward larger size was not seen in all of the extinct side-branches, some of which actually reversed direction and became smaller. Even the revised picture of horse evolution is oversimplified. Among other things, *Miohippus* actually appears in the fossil record before *Mesohippus,* though it persists after it.

Despite having been revised, the picture of horse evolution still includes a line connecting *Hyracotherium* with its supposed descendants, all the way up to the modern horse. Ironically, this very Darwinian line of ancestor-descendant relationships still presents a problem for neo-Darwinists like Simpson, because it is as consistent with directed evolution as the linear series in the old icon. The mere existence of extinct side-branches doesn't rule out the possibility that the evolution of modern horses was directed. A cattle drive has a planned destination, even though some steers might stray from the herd along the way. Or, to use another analogy, the branching pattern of arteries and veins in

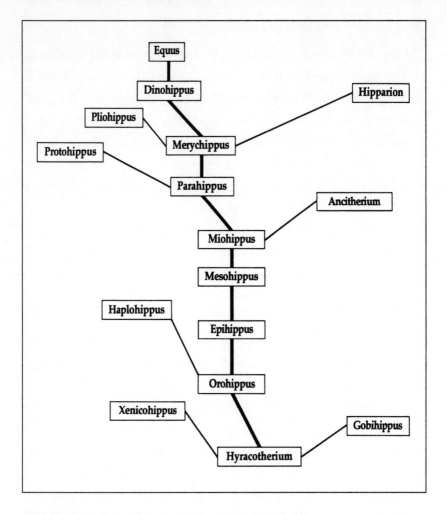

FIGURE 10-2 The new icon of horse evolution.

Two of the fossils shown in the old version, *Hyracotherium and Mesohippus,* are still considered to be in the line leading to modern horses, but *Protorohippus* has been dropped, and *Protohippus* is regarded as an extinct side-branch. Only a few of the many other extinct side-branches are shown here. Note that although the new pattern is not linear, it still shows a continuous lineage connecting *Hyracotherium* with the modern horse (heavy line).

the human body has some randomness to it, but our very lives depend on the fact that the overall pattern is predetermined.

This doesn't prove that directed evolution is true, but only that a branching-tree pattern in the fossil record doesn't refute it. A straight line and a branching tree are equally consistent (or inconsistent) with the existence (or non-existence) of either a predetermined goal or an inherent directive mechanism. In other words, even if we knew for sure what the pattern was, that alone would not be sufficient to establish whether or not horse evolution was directed.

What does the evidence really show?

Although the fossil pattern, by itself, does not refute directed evolution, it does seem to refute orthogenesis—if orthogenesis is taken to imply a straight line with no branches. But in the process of criticizing orthogenesis, Simpson made it clear that there was more at stake than straight-line evolution.

One thing at stake was the theory of inner forces or constraints. A mechanism was needed, and neo-Darwinists succeeded in persuading most biologists that theirs was the best—if not the only—candidate. But Simpson was criticizing even more than straight-line evolution and internal forces or constraints. By tacking "cosmic goals" onto the theory he was attacking, Simpson tried to strike a blow against the idea that evolution tends to follow some sort of pre-established plan.

If the whole of evolution were really the product of natural selection acting on random mutations, as neo-Darwinists claim, perhaps it would be legitimate to conclude that evolution is undirected in this cosmic sense. If peppered moths and Darwin's finches are our best evidence for natural selection, however, and the four-winged fruit fly is our best example of a morphological mutation, then neo-Darwinists are very far from proving their case. They don't have anywhere near enough evidence.

But the rejection of goal-directed evolution was around long before the fossil record of horses was revised, and long before neo-Darwinists proposed random genetic mutations and natural selection as the mechanism of evolutionary change. In fact, it was around before Othniel Marsh drew his picture of straight-line horse evolution in the 1880s.

Undirected evolution from Darwin to Dawkins

In Charles Darwin's view, the process of evolution by natural selection excluded designed results. He wrote: "There seems to be no more design in the variability of organic beings, and in the action of natural selection, than in the course which the wind blows." Darwin did not exclude design entirely, since the laws of nature—including the law of natural selection—might have been supernaturally designed. But he believed that survival of the fittest, acting on random variations, was inherently undirected, and thus could not produce designed results. He wrote that he was "inclined to look at everything as resulting from designed laws, with the details, whether good or bad, left to the working out of chance."

Darwin's view that evolution was undirected was not inferred from biological evidence. Natural selection had not yet been directly observed, and the nature and origin of variations was unknown. According to historian of science Neal Gillespie, Darwin excluded directed evolution and designed results because he wanted to place science on a foundation of materialistic philosophy. Since Darwin's view was primarily a philosophical doctrine rather than an empirical inference, its success depended less on marshalling evidence than on winning a war of ideas.

Simpson's rejection of directed evolution, like Darwin's, was a philosophical move rather than a scientific one. As Simpson put it, he favored the view that evolution "is dependent only on the physical possibilities of the situation and on the interplay of organism and environment, the usual materialist hypothesis." And he didn't limit himself to horses. Although the evidence for human evolution was (and still is) much scantier than that for horses, Simpson extrapolated his materialistic conclusion to our own species. "Man," he declared, "is the result of a purposeless and natural process that did not have him in mind."

Simpson wrote in the 1940s and 1950s, before Watson and Crick's discovery of the structure of DNA led to our current understanding of mutations as molecular accidents. By 1970 it seemed to many biologists that DNA mutations are the ultimate source of Darwin's random variations, and this seemed to confirm that evolution was undirected. When Jacques Monod announced in 1970 that "the mechanism of Darwinism is at last securely founded," he also declared: "And man has to understand that he is a mere accident."

Yet when Monod said this, the only beneficial DNA mutations known to him were biochemical. There was no evidence in 1970 that DNA mutations—random or not—could provide raw materials for morphological evolution. In other words, Monod—like Darwin and Simpson—was going far beyond the evidence in claiming that human beings are "a mere accident." Once again, the claim was philosophical rather than empirical.

This tendency to promote materialistic philosophy in the guise of biological science has continued. Oxford zoologist Richard Dawkins, as dogmatic a Darwinist as one might expect to find, is an outspoken apostle of what he calls "the blind watchmaker."

The blind watchmaker

Richard Dawkins's views on design in living things and direction in evolution are expressed most clearly in his 1986 book, *The Blind Watchmaker*. The book got its name from an argument made famous in the early nineteenth century by William Paley. "In crossing a heath," Paley wrote in 1802, "suppose I pitched my foot against a stone, and were asked how the stone came to be there." Paley answered that for all he knew, the stone might have been there forever. "But suppose I had found a watch upon the ground," Paley continued. Like any reasonable person, he would say that the watch had been made by a watchmaker.

For Paley, living things were like watches in their complexity and adaptiveness, so he argued that they must be designed. For Charles Darwin and Richard Dawkins, however, living things only *appear* to be designed. In fact, Dawkins defines biology as "the study of complicated things that give the appearance of having been designed for a purpose."

How does Dawkins know that design in living things is only apparent? Because, he says, natural selection explains all the adaptive features of living things, and natural selection is undirected. "Natural selection, the blind, unconscious, automatic process which Darwin discovered, and which we now know is the explanation for the existence and apparently purposeful form of all life, has no purpose in mind.... it is the *blind* watchmaker."

Although the subtitle of Dawkins's book is "Why the evidence of evolution reveals a world without design," it turns out that he actually excludes design on philosophical grounds. As he writes in his preface: "I want to persuade the reader, not just that the Darwinian world-view *happens* to be true, but that it is the only known theory that *could*, in principle, solve the mystery of our existence." And he repeats this claim in his concluding chapter: "Darwinism is the only known theory that is in principle *capable* of explaining certain aspects of life." (emphases in the original)

But claiming that a theory is true "in principle" is the hallmark of a philosophical argument, not a scientific inference. The latter requires evidence, and as Dawkins himself admits, evidence is unnecessary to prove the truth of Darwinism.

If Dawkins *were* making a scientific inference, he would have to have better evidence than computer simulations (the main "evidence" he provides in his book). He would need real evidence from living things. Yet, as we have seen throughout the preceding chapters, the real evidence for Darwin's theory is surprisingly thin. It appears to be overwhelming only because it is greatly exaggerated and sometimes blatantly misrepresented by certain proponents of Darwinian evolution. If there is anything about living things that is mere appearance, it is the alleged "evidence" that natural selection explains the existence and form of all life.

So Dawkins's exclusion of design and purpose is philosophical, not empirical. This is obvious not only from the insufficiency of the evidence, but also from the "in principle" form of his argument. It is also clear from the motivation that apparently underlies it. As Dawkins states early in his book, "Darwin made it possible to be an intellectually fulfilled atheist."

Now, Professor Dawkins has a right to profess atheism. He even has a right to make it intellectually fulfilling. But atheism is not science.

Teaching materialistic philosophy in the guise of science

There is nothing wrong with having philosophical views. Everyone does, whether they admit it or not. In public education, however, there is a reasonable expectation that philosophy be clearly identified as such, and not disguised as science. Certainly no philosophical view of human nature should be taught as though it were on a par with Newtonian physics or Mendelian genetics. Yet that is exactly what American public schools are doing in biology classrooms.

As we have seen, the doctrine that evolution was undirected, and consequently that human existence is a mere accident, is rooted in materialistic philosophy rather than empirical science. The doctrine existed long before the meager evidence now cited to justify it. Since the doctrine is very influential in our culture, it is a good idea to teach students about it—but as philosophy, not science.

Yet Miller and Levine's high school textbook, *Biology,* teaches students that as they learn about "the nature of life" they must "keep this concept in mind: *Evolution is random and undirected.*" (emphasis in the original) College students using *Life: The Science of Biology*, by Purves, Orians, Heller and Sadava, read that the Darwinian world view "means accepting not only the processes of evolution, but also the view that... evolutionary change is not directed toward a final goal or state."

Campbell, Reece and Mitchell's *Biology* treats students to an interview with Richard Dawkins, who tells them: "Natural selection is a bewilderingly simple idea. And yet what it explains is the whole of life, the diversity of life, the complexity of life, the apparent design of life," including human beings, who "are fundamentally not exceptional because we came from the same evolutionary source as every other species. It is natural selection of selfish genes that has given us our bodies and our brains." But our existence was not planned, because natural selection is the blind watchmaker, "totally blind to the future."

Students who have moved beyond introductory biology to study evolution in greater detail might find themselves reading Douglas Futuyma's textbook, *Evolutionary Biology.* According to Futuyma, Darwin's "theory of random, purposeless variations acted on by blind, purposeless natural selection provided a revolutionary new answer to almost all questions that begin with 'Why?'" The "profound, and deeply unsettling, implication of this purely mechanical, material explanation for the existence and characteristics of diverse organisms is that *we need not invoke, nor can we find any evidence for, any design, goal, or purpose anywhere in the natural world,* except in human behavior." (emphasis in original) Futuyma goes on to explain that "it was Darwin's theory of evolution, followed by Marx's materialistic (even if inadequate or wrong) theory of history and society and Freud's attribution of human behavior to influences over which we have little control, that provided a crucial plank to the platform of mechanism and materialism" that has since been "the stage of most Western thought."

Clearly, biology students are being taught materialistic philosophy in the guise of empirical science. Whatever one may think of materialistic philosophy, there is no doubt that it is being imposed on the evidence rather than inferred from it. And this is the real significance of neo-Darwinian efforts to revise the picture of horse evolution. Although there are scientific issues involved, what really matters is the myth.

CHAPTER 11

From Ape to Human: The Ultimate Icon

The most controversial aspect of Darwin's theory has always been its implications for human origins. Perhaps for this reason, Darwin did not even mention human evolution in *The Origin of Species*, except as a brief afterthought: "Much light will be thrown on the origin of man and his history." Twelve years went by before he wrote about this issue in any detail—in the first half of *The Descent of Man and Selection in Relation to Sex.*

According to Darwin, the origin of the human species was fundamentally similar to the origin of every other species. Human beings, he argued, are modified descendants of an ancestor they shared with other animals (most recently, the apes), and their distinctive features are due primarily (though not exclusively) to natural selection acting on small variations. Darwin's view had two implications which were (and continue to be) especially controversial: humans are nothing but animals, and they are not the preordained goal of a directed process.

But in Darwin's lifetime the evidence in favor of his theory was much too meager to support such sweeping claims about human nature. As far as Darwin knew, fossil evidence for human evolution had not yet been found, there was as yet no direct

FIGURE 11-1 The ultimate icon.

In a typical illustration of Darwin's theory of human origins, an ape-like creature is shown evolving through a series of hypothetical intermediate forms into a modern human.

FIGURE 11-1 The ultimate icon.

Although it is widely used to show that we are just animals, and that our very existence is a mere accident, the ultimate icon goes far beyond the evidence. Such drawings are (in Stephen Jay Gould's words) "incarnations of concepts masquerading as neutral descriptions of nature."

evidence for natural selection, and the origin of variations was unknown.

Despite the lack of evidence, the Darwinian view of human origins was soon enshrined in drawings that showed a knuckle-walking ape evolving through a series of intermediate forms into an upright human being. (Figure 11-1) Such drawings have subsequently appeared in countless textbooks, museum exhibits, magazine articles, and even cartoons. They constitute the ultimate icon of evolution, because they symbolize the implications of Darwin's theory for the ultimate meaning of human existence.

In the twentieth century, the ultimate icon seemed to acquire the evidence it initially lacked. Numerous fossil discoveries supplied what appeared to be transitional links in the evolutionary chain leading to modern humans; experiments on peppered moths and other organisms seemed to provide the missing evidence for natural selection; and geneticists thought they had found the raw materials for evolution in DNA mutations.

Yet the evidence is not as straightforward as it appears. As we have seen, Kettlewell's peppered moth experiments were flawed, and the oscillating natural selection observed in Darwin's finches produces no long-term evolution. Furthermore, although beneficial DNA mutations occur at the biochemical level, the widely advertised morphological mutations in four-winged fruit flies produce cripples, not raw materials for evolution.

Finally, as we shall see in this chapter, interpretations of the fossil evidence for human evolution are heavily influenced by personal beliefs and prejudices. Experts in paleoanthropology—the study of human origins—acknowledge that their field is the most subjective and contentious in all of biology—hardly a firm foundation for the far-reaching claims some Darwinists want to make about human nature.

Are we (just) animals?

Darwin began *The Descent of Man* by reminding readers that "man is constructed on the same general type or model as other mammals." After reviewing evidence for evolution that he had presented in *The Origin of Species*—especially the supposed similarities between the embryos of humans and other vertebrates—he concluded that "man bears in his bodily structure clear traces of his descent from some lower form."

"My object," Darwin explained, "is to show that there is no fundamental difference between man and the higher animals in their mental faculties." He argued that all have "similar passions, affections, and emotions, even the more complex ones, such as jealousy, suspicion, emulation, gratitude, and magnanimity... they possess the same faculties of imitation, attention, deliberation, choice, memory, imagination, the association of ideas, and reason, though in very different degrees." Thus "the difference in mind between man and the higher animals, great as it is, certainly is one of degree and not of kind."

For Darwin, the continuity between animals and humans extended even to morality and religion. It seemed to him that "any animal whatever, endowed with well-marked social instincts, the parental and filial affections being here included, would inevitably acquire a moral sense or conscience, as soon as its intellectual powers had become as well, or nearly as well developed, as in man." And the "tendency in savages to imagine that natural objects and agencies are animated by spiritual and living essences," which Darwin compared to a dog's tendency to imagine hidden agency in things moved by the wind, "would easily pass into the belief in the existence of one or more gods." Thus the "feeling of religious devotion" is merely a higher form of "the deep love of a dog for his master."

There are at least three questions here. First, do human beings have some features in common with other animals? Second, did human beings acquire these features through descent with modification from animal ancestors? And third, are humans *just* animals? Darwin explicitly answered "yes" to the first two questions; and by maintaining that human morality and religion differ only in degree rather than kind from animal instincts, he implicitly answered "yes" to the third.

Some modern Darwinists write as though it was Darwin who showed us that we are part of the natural world. For example, Oxford zoologist Richard Dawkins wrote in 1989 that Darwin shocked "the vanity of our species" by showing that we are "close cousins to… monkeys and apes," thus proving that "we too are animals."

But the awareness that the human body is part of nature was around long before Darwin. It was affirmed by thirteenth-century Catholic theologian and philosopher Thomas Aquinas, who even included emotive responses among the features that humans share with other animals. And eighteenth-century creationist Carolus Linnaeus, who devised the modern system of biological classification, placed humans in the primate order with apes and monkeys. In other words, by answering "yes" to the first question Darwin wasn't saying anything new.

Of course, the tradition represented by Aquinas maintained that human beings have a spiritual nature as well as an animal one. When Darwin implicitly answered "yes" to the third question, and claimed that human beings are *nothing more* than animals, he departed from this tradition. Even here, however, Darwin wasn't saying anything new. Materialistic philosophers since ancient Greece had been saying the same thing.

Darwin's novel contribution was to claim that descent with modification accounted for all of human nature, including the part previously attributed to spirit. He thereby provided materialistic philosophy with what appeared to be scientific support. But before Darwin's claim could qualify as science rather than philosophy, it required evidence.

Finding evidence to fit the theory

Although "Neanderthal Man" had been discovered in 1856, he was not then regarded as an ancestor of human beings. According to one popular theory, his bones were different from those of a modern human because they had been deformed by disease. In any case, Darwin and his immediate followers had to argue for their theory without any fossil evidence for human evolution.

In the absence of fossil evidence, similarities between humans and living apes served as a proxy. In an 1863 book entitled *Evidence as to Man's Place in Nature,* Thomas Henry Huxley compared skeletons of apes to that of a human to show the gradations between them. (Figure 11-2) "But if Man be separated by no greater structural barrier from the brutes than they are from one another," wrote Huxley, "then, there would be no rational ground for doubting that man might have originated... by the gradual modification of a man-like ape [or] as a ramifica-

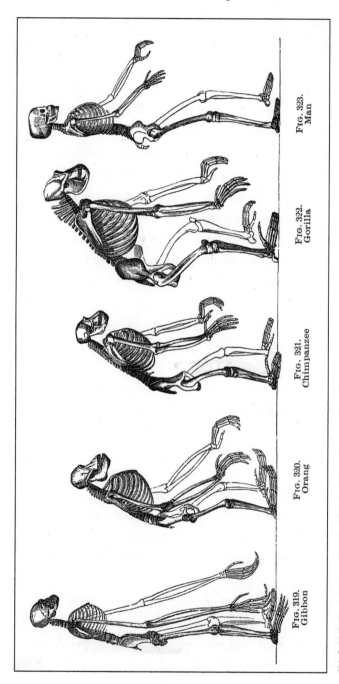

FIG. 319.
Gibbon

FIG. 320.
Orang

FIG. 321.
Chimpanzee

FIG. 322.
Gorilla

FIG. 323.
Man

FIGURE 11-2 Huxley's version of the ultimate icon.

Skeletons of a gibbon, orangutan, chimpanzee, gorilla, and human, arranged in a series showing progression toward the human form. From Thomas Henry Huxley's 1863 book, *Evidence as to Man's Place in Nature.*

tion of the same primitive stock as those apes." Huxley concluded: "Man is, in substance and in structure, one with the brutes."

The striking similarity between Huxley's illustration and the ultimate icon is unmistakable. Yet neither Huxley nor Darwin believed that living apes were our ancestors. What Huxley's illustration shows is that, from the very beginning, the ape-to-human icon was simply a restatement of materialistic philosophy. Its form preceded any fossil evidence of ancestor-descendant relationships, and it made do with whatever evidence happened to be at hand—in this case, similarities to living apes. Fossils discovered later were just plugged into this preexisting framework.

Neanderthal was not initially among them. Huxley knew about Neanderthal, but like most of his contemporaries he regarded it as fully human, rather than ancestral to humans. A few decades later, however, after more fossils had been found, French paleontologist Marcellin Boule declared that Neanderthal was *not* human, and not even ancestral to humans. Instead, he regarded it as an extinct side branch of the evolutionary tree.

According to Boule, Neanderthals had a stooped posture, midway between apes and humans—the "cave man" image subsequently immortalized in countless cartoons. Paleoanthropologists are now convinced that Boule was wrong, and that Neanderthals walked upright just as we do. But this realization came later; in the early twentieth century most people accepted Boule's interpretation, and excluded Neanderthals from the evolutionary line leading to human beings.

Without Neanderthal, however, there was still no fossil evidence for human origins. Where were the ancestors required by Darwin's theory? Dutch anatomist Eugene Dubois had found some fossil bones in Java in the 1890s, but his claim that "Java Man" was intermediate between apes and humans was widely disputed. It wasn't until 1912 that amateur paleontologist Charles Dawson announced that he had found what everyone was looking for, in a gravel pit at Piltdown, England.

The Piltdown fraud

Dawson had found some pieces of human skull and part of an ape-like lower jaw with two teeth. He took them to Arthur Smith Woodward at the British Museum, who reconstructed an entire skull from the fragments and reported the discovery to the Geological Society of London in December 1912. Although some paleontologists were skeptical, subsequent discoveries at the same site seemed to confirm Smith Woodward's conclusion that "Dawson's Dawn Man" was the missing link needed to confirm evolutionary theory.

That theory, as understood in 1912, predicted that the ancestor of human beings would have a large brain and an ape-like jaw. The Piltdown specimen fit the prediction so well that nobody checked closely to determine whether the skull and jaw fragments belonged to the same individual. Smith Woodward's reconstruction was at first disputed, but then widely accepted, and for several decades all newly discovered fossils were interpreted in the light of "Piltdown Man." Only after several fossils had been found that couldn't be shoehorned into the existing theory did ideas about human origins begin to change. Then, having already lost much of its iconic status, Piltdown was exposed as a fraud.

In 1953 Joseph Weiner, Kenneth Oakley, and Wilfrid Le Gros Clark proved that the Piltdown skull, though perhaps thousands of years old, belonged to a modern human, while the jaw fragment was more recent, and belonged to a modern orangutan. The jaw had been chemically treated to make it look like a fossil, and its teeth had been deliberately filed down to make them look human. Weiner and his colleagues concluded that Piltdown man was a forgery.

Most modern biology textbooks do not even mention Piltdown. When critics of Darwinism bring it up, they are usually told that the incident merely proves that science is self-correcting. And so it was, in this case—though the self-correcting took over forty years. But the more interesting lesson to be learned from Piltdown is that scientists, like everyone else, can be fooled into seeing what they want to see.

The features that pointed to fraud in 1953 had been there all along. As paleoanthropologist Roger Lewin wrote recently: "Given all the many anatomical incongruities in the Piltdown remains, which of course are glaringly obvious from the vantage of the present, it is truly astonishing that the forgery was so eagerly embraced." Thus "the real interest of Piltdown" is "how those who believed in the fossil saw in it what they wanted to see." And according to historian of biology Jane Maienschein, Piltdown shows us "how easily susceptible researchers can be manipulated into believing that they have actually found just what it was they had been looking for."

Many human-like fossils have been found since 1912, and unlike Piltdown they appear to be genuine. Some have distinctively ape-like features, while others are more human-like. But even genuine fossils that bear on human origins have typically been so controversial that in 1970 British anthropologist John Napier called them "bones of contention." And each new discovery seems to add to the problem rather than alleviate it. In 1982 American paleontologists Niles Eldredge and Ian Tattersall noted that it is a "myth that the evolutionary histories of living things are essentially a matter of discovery." If this were really true, they wrote, "one could confidently expect that as more hominid fossils were found the story of human evolution would become clearer. Whereas if anything, the opposite has occurred."

There are at least two reasons for this. One is that the fossil evidence leaves a lot of room for interpretation. The other is that the subjectivity that prepared the way for Piltdown continues to plague human origins research.

Trying to fit every new peg into the same size hole

How much can the fossils show us?

The fossil evidence is open to many interpretations because individual specimens can be reconstructed in a variety of ways, and because the fossil record cannot establish ancestor-descendant relationships.

One famous fossil skull, discovered in 1972 in northern Kenya, changed its appearance dramatically depending on how the upper jaw was connected to the rest of the cranium. Roger Lewin recounts an occasion when paleoanthropologists Alan Walker, Michael Day, and Richard Leakey were studying the two sections of "skull 1470." According to Lewin, Walker said: "You could hold the [upper jaw] forward, and give it a long face, or you could tuck it in, making the face short.... How you held it really depended on your preconceptions. It was very interesting watching what people did with it." Lewin reports that Leakey recalled the incident, too: "Yes. If you held it one way, it looked like one thing; if you held it another, it looked like something else."

Just recently, *National Geographic* magazine commissioned four artists to reconstruct a female figure from casts of seven fossil bones thought to be from the same species as skull 1470. One artist drew a creature whose forehead is missing and whose jaws look vaguely like those of a beaked dinosaur. Another artist drew a rather good-looking modern African-American woman with unusually long arms. A third drew a somewhat scrawny female with arms like a gorilla and a face like a Hollywood werewolf. And a fourth drew a figure covered with body hair and climbing a tree, with beady eyes that glare out from under a heavy, gorilla-like brow.

This remarkable set of drawings shows clearly how a single set of fossil bones can be reconstructed in a variety of ways. Someone looking for an intermediate form to plug into an ape-to-human sequence could pick whichever drawing seems to fit best. (Not surprisingly, the strongly pro-Darwin *National Geographic* buried these revealing drawings on an unnumbered page among the advertisements at the back of the magazine.)

Another reason why fossils have not solved the problem of human origins is the difficulty or impossibility of determining ancestor-descendant relationships from the fossil record. In 1981 Constance Holden wrote in *Science:* "The primary scientific evidence is a pitifully small array of bones from which to construct man's evolutionary history. One anthropologist has compared the task to that of reconstructing the plot of *War and Peace* with 13 randomly selected pages."

Henry Gee, Chief Science Writer for *Nature*, is even more pessimistic. "No fossil is buried with its birth certificate," he wrote in 1999, and "the intervals of time that separate fossils are so huge that we cannot say anything definite about their possible connection through ancestry and descent." It's hard enough, with written records, to trace a human lineage back a few hundred years. When we have only a fragmentary fossil record, and we're dealing with millions of years—what Gee calls "Deep Time"—the job is effectively impossible.

Gee regards each fossil as "an isolated point, with no knowable connection to any other given fossil, and all float around in an overwhelming sea of gaps." He points out, for example, that all the evidence for human evolution "between about 10 and 5 million years ago—several thousand generations of living creatures—can be fitted into a small box." Thus the conventional picture of human evolution as lines of ancestry and descent is "a completely human invention created after the fact, shaped to accord with human prejudices." Putting it even more bluntly, Gee concludes: "To take a line of fossils and claim that they represent a lineage is not a scientific hypothesis that can be tested, but an assertion that carries the same validity as a bedtime story—amusing, perhaps even instructive, but not scientific."

If individual fossils lend themselves to such varied interpretations, however, and evolutionary history cannot be reconstructed from the fossil record, where do stories of human evolution come from?

Paleoanthropology: science or myth?

At a meeting of the British Association for the Advancement of Science in the early 1980s, Oxford historian John Durant asked: "Could it be that, like 'primitive' myths, theories of human evolution reinforce the value-systems of their creators by reflecting historically their image of themselves and of the society in which they live?" Durant later wrote that "it is surely worth asking whether ideas about human evolution might serve essentially similar functions in both pre-scientific and scientific cultures…. Time and again, ideas of human origins turn out on closer examination to tell us as much about the present as the past, and as much about our own experiences as about those of our remote ancestors." Durant concluded: "As things stand at the present time, we are in urgent need of the de-mythologisation of science."

A few years later, Duke University anthropologist Matt Cartmill told a meeting of the American Association of Physical Anthropologists that some aspects of their science lay "within the province of ideology and religion, broadly defined." As reported by science writer Roger Lewin, many anthropologists reacted to this with something like the following: "Well, I guess in the early days people's work used to be affected by this sort of thing—ideology, mythology, and so on—but not now; not now that anthropology is *really* scientific." (emphasis in the original) Cartmill's response was unyielding: "This tendency to rescue scientific appearances by evading the mythological point of our science has distorted paleoanthropological thought through most of the twentieth century."

At Yale Graduate School in the late 1970s, paleoanthropologist Misia Landau was struck by the similarity between accounts of human evolution and old-fashioned folk tales. In a 1991 book on the subject, *Narratives of Human Evolution*, she maintained that many "classic texts in paleoanthropology" were "determined as much by traditional narrative frameworks as by material evidence." The typical framework was that of a folktale in which a hero (i.e., our ancestor) leaves a relatively safe haven in the trees, sets out on a dangerous journey, acquires various gifts, survives a series of tests, and is finally transformed into a true human being.

According to Landau, when paleoanthropologists want to explain what really happened in human evolution they use four main events. These are: moving from trees to the ground, developing upright posture, acquiring intelligence and language, and developing technology and society. Although Landau found these four elements in all accounts of human evolution, their order varied depending on the viewpoint of the narrator. She concluded that "themes found in recent paleoanthropological writing... far exceed what can be inferred from the study of fossils alone and in fact place a heavy burden of interpretation on the fossil record—a burden which is relieved by placing fossils into preexisting narrative structures." <u>Paleoanthropologists, in other words, are storytellers.</u> *all historians are...*

The mythical elements in the study of human origins are still there. In 1996 American Museum of Natural History Curator Ian Tattersall acknowledged that "in paleoanthropology, the patterns we perceive are as likely to result from our unconscious mindsets as from the evidence itself." Arizona State University anthropologist Geoffrey Clark echoed this view in 1997 when he wrote that "we select among alternative sets of research conclusions in accordance with our biases and preconceptions—a process that is, at once, both political and subjective." Clark suggested "that paleoanthropology has the form but not the substance of a science."

Given the highly subjective nature of paleoanthropology—as acknowledged by its own practitioners—what can the field reliably tell us about human origins?

What do we know about human origins?

Obviously, the human species has a history. Many fossils have been found that appear to be genuine, and many of them have some features that are ape-like and some that are human-like. On these statements, all paleoanthropologists would no doubt agree.

When it comes to reconstructing entire individuals or the history of human evolution, however, agreement is hard to find. One area of disagreement is how many species of human-like apes or ape-like humans co-existed at any given moment. The "lumpers" tend to group all specimens into one or a few species, while the "splitters" divide them into many more. Even if agreement were to be reached on which specimens represent separate species, there would still be the question of whether they are ancestors of modern humans or extinct side-branches of the evolutionary tree. Disagreement also continues between the "Out of Africa" camp, which maintains that modern humans first evolved in Africa and then spread throughout the world, and the "Multiregional" camp, which argues that our species evolved in many places simultaneously.

Currently in the news is the never-ending controversy over Neanderthals. Were they our ancestors? Were they a separate species, now extinct? Or were they a race of humans, eventually absorbed into our modern global family? Almost every month, a proponent of one view or another takes to the print media or the airwaves, declaring the matter settled. Wait a few months, however, and someone will probably say the opposite with equal confidence. In 1995 science writer James Shreeve reported that he had "talked to one hundred and fifty scientists—archaeologists, anatomists, geneticists, geologists, dating experts—and sometimes it seemed I had come away with one hundred and fifty different points of view" about the place of Neanderthals in human evolution. Any theory about Neanderthals is like the weather in many parts of the country: If you don't like it, wait a little while and it will change.

Anyone who follows these controversies for any length of time is likely to become somewhat cynical about the prospects for resolving them. In 1996 Berkeley evolutionary biologist F. Clark Howell wrote: "There is no encompassing theory of [human] evolution… Alas, there never really has been." The field is characterized by "narrative treatments" based on little evidence, so "it is probably true that an encompassing scenario" of human evolution "is beyond our grasp, now if not forever."

Howell's pessimism was echoed by Arizona State University anthropologist Geoffrey Clark in 1997: "Scientists have been trying to arrive at a consensus about modern human origins for more than a century. Why haven't they been successful?" In Clark's opinion, it is because paleoanthropologists proceed from such different "biases, preconceptions and assumptions." Thus explanatory models of human evolution, according to Clark, "are little more than a house of cards—remove one card… and the whole structure of inference is threatened with collapse."

The general public is rarely informed of the deep-seated uncertainty about human origins that is reflected in these statements by scientific experts. Instead, we are simply fed the latest version of somebody's theory, without being told that paleoanthropologists themselves cannot agree over it. And typically, the theory is illustrated with fanciful drawings of cave men, or human actors wearing heavy makeup.

Add to these visual effects some "just-so" stories about the hypothetical adaptive value of descending from the trees, or of learning how to use tools, or of switching from hunting to agriculture, and the account is complete. Popular presentations of this sort can be found in the "Dawn of Humans" series in *National Geographic* magazine, occasional cover stories in *Time* or *Newsweek*, and periodic television specials on the Discovery Channel. Such presentations typically mention a few minor disagreements among paleoanthropologists, but the public is rarely told that the fossils have been placed into "preexisting narrative structures" or that the story they are hearing rests on "biases, preconceptions and assumptions." It seems that never in the field of science have so many based so much on so little.

Woven into the mythical accounts of human evolution is usually the message that we are nothing more than animals. Yet the message was around long before the meager evidence that is now plugged into the narratives to make them sound scientific. Whether the ultimate icon is presented in the form of a picture or a narrative, it is old-fashioned materialistic philosophy disguised as modern empirical science.

And the claim that humans are mere animals is not the only philosophical pill we are expected to swallow. Since the 1970s, the ultimate icon has increasingly been used to promote the doctrine that evolution was undirected, and that our existence is a mere accident.

Concepts masquerading as neutral descriptions of nature

One of the most vocal critics of directed evolution has been Harvard paleontologist Stephen Jay Gould. In fact, the epigraph that introduces this book was taken from Gould's critique of "the iconography of progress" in his 1989 book, *Wonderful Life.* When Gould alerts his readers to "the evocative power of a well-chosen picture," and warns them that "ideas passing as descriptions lead us to equate the tentative with the unambiguously factual," his eloquence is aimed at the idea of goal-oriented evolution.

As might be expected, Gould rejects the old "ladder of progress" image that Simpson had found unacceptable in the idea of orthogenesis. Surprisingly, however, Gould also rejects the branching-tree pattern which Simpson put in its place. Gould calls Darwin's branching tree the "cone of increasing diversity," and argues that it misrepresents the history of life. That history, according to Gould, is characterized by maximal diversity early on (in the Cambrian explosion), followed later by "decimation" as various lineages become extinct. So Gould replaces both the ladder and the cone of increasing diversity with the "iconography of decimation."

Gould argues that the fact of extinction is the most powerful antidote to the poisonous idea of progress. In his view, extinctions are accidents that demonstrate the fundamental "contingency" of evolution. If we could "replay the tape" of life's history, we would find that it never tells the same story twice. The contingency and irreproducibility of evolution destroy any notion of "human inevitability and superiority," and teach us that we are mere accidents.

But how does Gould know that extinctions are accidents? On the basis of fossil evidence, how could he possibly know? Clearly, it takes more than a pattern in the fossil record to answer sweeping questions about direction and purpose—even if we knew for sure what those patterns are. And even if extinctions are accidents, does that rule out the possibility that evolution is goal-oriented? Everyone's death is contingent; does that make everyone's birth and life an accident? The continued existence of the human species is contingent on many things: That we don't blow ourselves up with nuclear weapons, that the earth isn't struck by a large asteroid, and that we don't poison our environment, among other things. But it doesn't follow that our very existence is an accident, or that human life is purposeless.

Canadian philosopher of biology Michael Ruse recently criticized the tendency of Gould and others to use biological evolution as a platform for sermonizing about the meaning of human existence. "If people want to make a religion of evolution, that is their business," Ruse wrote, but "we should recognize when people are going beyond the strict science, moving into moral and social claims, thinking of their theory as an all-embracing world picture. All too often, there is a slide from science to something more."

Ruse is what might be called a moderate or self-critical Darwinist. He calls himself "an ardent evolutionist," yet he objects when "evolution is promoted by its practitioners as more than mere science. Evolution is promulgated as an ideology, a secular religion."

So Gould's sermons on contingency, like the materialistic views of Darwin, Huxley, Simpson, Monod, and Dawkins, are based on personal philosophy, not empirical evidence. Although Gould has the same right as everyone else to express his views, they should not be taught as though they were science. Nevertheless, like the philosophical views of Richard Dawkins, Gould's are now featured in some biology textbooks. Raven and Johnson's 1999 *Biology* includes an interview with Gould, who declares: "Humans represent just one tiny, largely fortuitous, and late-arising twig on the enormously arborescent bush of life."

Like so many other things we have encountered, this is not science, but myth.

> How is this a myth?

CHAPTER 12

Science or Myth?

"No educated person any longer questions the validity of the so-called theory of evolution, which we now know to be a simple fact" announced Ernst Mayr in the July 2000 issue of *Scientific American*. Mayr continued: "Likewise, most of Darwin's particular theses have been fully confirmed, such as that of common descent, the gradualism of evolution, and his explanatory theory of natural selection."

Ask any educated person how we know that evolution is a simple fact, and that Darwin's particular theses have been fully confirmed, and chances are that person will list some or all of the icons described in this book. For most people—including most biologists—the icons *are* the evidence for Darwinian evolution.

As we have seen, however, the icons of evolution *misrepresent* the evidence. One icon (the Miller-Urey experiment) gives the false impression that scientists have demonstrated an important first step in the origin of life. One (the four-winged fruit fly) is portrayed as though it were raw materials for evolution, but it is actually a hopeless cripple—an evolutionary dead end. Three icons (vertebrate limbs, *Archaeopteryx*, and Darwin's finches) show actual evidence but are typically used to conceal fundamental problems in its interpretation. Three (the tree of life, fossil horses, and human origins) are incarnations of concepts masquerading as neutral descriptions of nature. And two icons (Haeckel's embryos, and peppered moths on tree trunks) are fakes.

People such as Ernst Mayr insist that there is overwhelming evidence for Darwin's theory. But the icons of evolution have been advertised for years as the best evidence we have. Even most evolutionary biologists think so. After all, until very recently Douglas Futuyma did not doubt Haeckel's embryos, and Jerry Coyne did not doubt peppered moths. If there is such overwhelming evidence for Darwinian evolution, why do our biology textbooks, science magazines and television nature documentaries keep recycling the same tired old myths?

There is a pattern here, and it demands an explanation. Instead of continually testing their theory against the evidence, as scientists are supposed to do, some Darwinists consistently ignore, explain away, or misrepresent the biological facts in order to promote their theory. One isolated example of such behavior might be due simply to overzealousness. Maybe even two. But ten? Year after year?

Before turning to the implications of this pattern, it is important to remind ourselves that most ordinary biologists have never noticed it. Most biologists are honest, hard-working scientists who insist on accurate presentation of the evidence, but who rarely venture outside of their own fields. The truth about the icons of evolution will surprise them as much as it surprises everyone else. Many of these biologists believe in Darwinian evolution because that's what they learned from their textbooks. In other words, they have been misled by the same misrepresentations that have fooled the general public.

These biologists suffer from the "specialist effect"—their expertise is limited to a particular field. A few years ago, Berkeley law professor and Darwin critic Phillip E. Johnson was discussing evolution with a well-known cell biologist. The biologist insisted that Darwinian evolution is generally true, but acknowledged that it could not explain the origin of the cell. "Has it occurred to you," Johnson said, "that the cell is the only thing you know anything about?"—suggesting that if he knew more about other fields he would realize that Darwinian evolution doesn't work in them, either. Thus it is with many biologists: They realize that Darwinian evolution cannot adequately explain what they know in their own field, but assume that it explains what they don't know in others.

So even though most biologists might consider themselves Darwinists, in many cases it is only because they believe what their more dogmatic colleagues are telling them. How about the dogmatists themselves? Can they also claim to be innocent victims of the specialist effect? Or is something else going on?

The "F" word

Fraud is a dirty word. In their 1982 book, *Betrayers of the Truth: Fraud and Deceit in the Halls of Science*, William Broad and Nicholas Wade distinguish between deliberate fraud and unwitting self-deception. Conscious faking of data is an example of the former, but is relatively rare. Unconscious manipulation of data by researchers convinced that they already know the truth is an example of the latter, and is much more common. There is a continuum between fraud and self-deception, and most cases of misrepresentation fall somewhere between them.

Some textbook-writers, such as Douglas Futuyma, may not even know that one or more of the icons of evolution are false.

Futuyma might reasonably be criticized for his ignorance—especially since he is supposed to be an expert on this subject—but ignorance is not conscious misrepresentation.

What about Stephen Jay Gould, a historian of science who has known for decades about Haeckel's faked embryo drawings? All that time, students passing through Gould's classes were learning biology from textbooks that probably used Haeckel's embryos as evidence for evolution. Yet Gould did nothing to correct the situation until another biologist complained about it in 1999. Even then, Gould blamed textbook-writers for the mistake, and dismissed the whistle-blower (a Lehigh University biochemist) as a "creationist." Who bears the greatest responsibility here—textbook-writers who mindlessly recycle faked drawings, people who complain about them, or the world-famous expert who watches smugly from the sidelines while his colleagues unwittingly become accessories to what he himself calls the "academic equivalent of murder"?

The revelation that the peppered moth story is flawed came only recently compared to the truth about Haeckel's embryos, so perhaps some textbook-writers can be excused for continuing to use it. Yet every biologist who works on peppered moths has known for over a decade that the moths don't rest on tree trunks, and that the textbook pictures have been staged. If science is self-correcting, why haven't the experts taken the initiative to get the faked photos out of the textbooks?

What about textbook-writers who *know* they are distorting the truth? As we saw in the chapter on peppered moths, Canadian Bob Ritter (assuming he was correctly quoted in the *Alberta Report Newsmagazine*) knowingly included staged pictures in his biology textbook. "How convoluted do you want to make it for a first time learner?" Ritter asked. "We want to get across the idea of selective adaptation." Ritter knew he was misrepresenting the truth, but defended his action on the grounds that he was illustrating a basic principle. Is it legitimate to illustrate a principle—even a true principle—with an icon known to be false? Do hidden convictions justify open falsehoods?

When paleontologists published the official description of *Bambiraptor* in March 2000, they decorated the animal with imaginary feathers. They knew that these structures had not been found with the fossil, yet the only indication of this in their publication was an obscure phrase in a figure caption. When a Chinese fossil dealer glues together two different skeletons to make them look like one animal, he is committing fraud. When paleontologists put feathers on a dinosaur to make it look like a bird, does an obscure disclaimer make their action much better?

These are difficult questions, with potentially serious consequences for biologists. What should be our guidelines in answering them?

Scientific misconduct and stock fraud

According to Harvard biologist Louis Guenin, U.S. securities laws provide "our richest source of experiential guidance" in defining what constitutes scientific misconduct. "The pivotal concept here is candour," wrote Guenin in *Nature* in 1999, "the attribute on a given occasion of not uttering anything that one believes false or misleading. We describe breaches of candour as deception." Guenin continued: "An investigator induces and betrays a listener's trust by signalling 'I believe it' while believing a false utterance false or a misleading omission misleading."

As we saw, the average beak size in one species of Darwin's finches increased 5 percent during a severe drought, and the authors of a National Academy of Sciences booklet claimed that "if droughts occur about once every ten years on the islands, a new species of finch might arise in only about 200 years." Yet the authors of the booklet omitted the fact that the average beak size returned to normal after the drought ended. Berkeley law professor Phillip E. Johnson called this "the sort of distortion that would land a stock promoter in jail."

If security laws provide our best guidance in determining scientific misconduct, the analogy is appropriate. A stock promoter who tells his clients that a particular stock can be expected to double in value in twenty years because it went up 5 percent in 1998, but conceals the fact that the same stock declined 5 percent in 1999, might well be charged with fraud. U.S. securities laws prescribe severe penalties for anyone who deliberately misstates or omits material facts in securities transactions.

What about scientists who knowingly make false utterances or misleading omissions but believe the overall effect is not misleading because they are teaching "a deeper truth"? Does the commitment to a supposed deeper truth excuse conscious misrepresentation? Such an excuse probably wouldn't help a stock promoter. Under federal law, a stock promoter is not justified in misstating the facts just because he or she deeply believes that a company is destined to prosper. The stock promoter commits fraud by misrepresenting the truth, regardless of his or her underlying beliefs. Shouldn't scientists be held to the same standard?

Fraud is a dirty word, and it should not be used lightly. In the cases described in this book, dogmatic promoters of Darwinism did not see themselves as deceivers. Yet they seriously distorted the evidence—often knowingly. If this is fraud when a stock promoter does it, what is it when a scientist does it?

Of course, there are differences between the stock market and the scientific enterprise. But science is the search for truth, so if anything it should be held to a higher standard than stock-trading. If the icons of evolution distort the truth, we should not be using them to teach biology to impressionable students. Yet some dogmatic Darwinists have exploited their evocative power to a degree that would make demagogues and advertising executives blush.

This is not what we have been led to expect from scientists. Although we are now accustomed to spin doctors in politics and advertising, we rightly hold scientists to a higher standard of honesty. The promoters of the icons of evolution style themselves as defenders of the truth, besieged (at least in America) by the dark forces of ignorance and religious fundamentalism. Apparently, they are not what they pretend to be.

If dogmatic promoters of Darwinian evolution were merely distorting the truth, that would be bad enough. But they haven't stopped there. They now dominate the biological sciences in the English-speaking world, and use their position of dominance to censor dissenting viewpoints.

Darwinian censorship

As we saw in Kevin Padian's "cracked kettle" approach to biology, dogmatic Darwinists begin by imposing a narrow interpretation on the evidence and declaring it to be the only way to do science. Critics are then labeled unscientific; their articles are rejected by mainstream journals, whose editorial boards are dominated by the dogmatists; the critics are denied funding by government agencies, who send grant proposals to the dogmatists for "peer" review; and eventually the critics are hounded out of the scientific community altogether.

In the process, evidence against the Darwinian view simply disappears, like witnesses against the Mob. Or the evidence is buried in specialized publications, where only a dedicated researcher can find it. Once critics have been silenced and counter-evidence has been buried, the dogmatists announce that there is no scientific debate about their theory, and no evidence against it. Using such tactics, defenders of Darwinian orthodoxy have managed to establish a near-monopoly over research grants, faculty appointments, and peer-reviewed journals in the United States.

In April 2000 a furor erupted at Baylor University in Texas over the right of academics to dissent from Darwinian orthodoxy. The Michael Polanyi Center, named after a noted philosopher of science, had been established six months earlier by the University administration to promote research on the conceptual foundations of science. When the Center sponsored a major international conference (numbering among its participants two Nobel laureates), all hell broke loose, because the faculty learned that the Center's director, William Dembski, was openly critical of Darwinian evolution.

The Baylor Faculty Senate immediately voted to shut down the Michael Polanyi Center, complaining that the university's president, Robert Sloan, had failed to get their approval before opening it. But Sloan pointed out that other centers had been created in the same way during and before his administration, and maintained that the real issue was whether "the old paradigms— Darwinism and neo-Darwinism—can be challenged." Professor Jay Losey, chair-elect of the Faculty Senate, confirmed Sloan's assessment: "If you dismiss or belittle evolution," he said, "then you call into question the whole endeavor of modern science." Baylor University spokesman Larry Brumley found it ironic that faculty members who claim to defend academic freedom were denying it in this case, and called their effort to close the Center "a form of censorship." Sloan said it "borders on McCarthyism." As of this writing, the future of the Michael Polanyi Center at Baylor is uncertain.

Dogmatic defenders of Darwinian evolution control not only most American universities, but they also wield enormous power over most public school systems. Kevin Padian is president of the ironically misnamed National Center for Science Education (NCSE), which pressures local school districts to prohibit classroom challenges to Darwinian evolution. (The executive director of the NCSE was a co-author of the National Academy's 1998 booklet on evolution that included the sort of distortion that would land a stock promoter in jail.) In 1999, when a school district near Detroit wanted to put some books critical of Darwinism in the high school library, the NCSE strongly advised them against it.

The NCSE tells school boards that "evolution isn't scientifically controversial," so "arguments against evolution" are "code words for an attempt to bring non-scientific, religious views into the science curriculum." Since U.S. courts have declared it unconstitutional to teach religion in public schools, this amounts to a warning that the school board is contemplating something illegal. If the warning doesn't work, the NCSE calls on the American Civil Liberties Union (ACLU) for backup, and the ACLU sends a letter to the school board threatening an expensive lawsuit. Since every school district in the country is already struggling to make ends meet, this bullying by the NCSE and ACLU has been quite successful in blocking overt criticism of Darwinian evolution in public school classrooms.

In Burlington, Washington, high-school biology teacher Roger DeHart taught evolution for years, but supplemented his pro-Darwinian textbook with material criticizing Darwinian evolution from the perspective of "intelligent design theory." In 1997 the ACLU wrote a letter to the local school board threatening legal action on the grounds that intelligent design theory is religious rather than scientific. DeHart withdrew the disputed materials, but requested permission to provide others dealing with scientific problems in Darwin's theory.

After extended negotiations, DeHart submitted for approval several articles from mainstream science publications. The articles question the scientific accuracy of Haeckel's embryos and the peppered moth story, both of which were presented uncritically in the textbook DeHart was required to use. In May 2000, under pressure from local ACLU members, Burlington school officials prohibited DeHart from using the articles. Despite its name, the ACLU did not object to this egregious act of censorship, apparently less concerned with defending civil liberties than with shielding Darwinian orthodoxy from criticism.

In 1999, when the Kansas State Board of Education was considering new statewide curricular standards, the strongly pro-Darwin members of a writing committee proposed a ninefold increase in the coverage of evolution compared to the 1995 standards. They demanded that biological evolution be made one of the "unifying concepts and processes" of science, on a par with such basic categories as "organization," "explanation," "measurement," and "function." They also wanted students to "understand" that large-scale evolutionary changes are explained by natural selection and genetic changes.

The Kansas Board increased the treatment of evolution fivefold over the previous standards, but rejected the writing committee's demand to install biological evolution as a unifying concept of science. Some Board members wanted to include the Darwinian explanation for large-scale evolution as long as students were exposed to evidence against it; but when pro-Darwin Board members refused to agree to this, the topic was omitted. Dissatisfied with the outcome, the Darwinists informed the major news media that the Board had eliminated evolution entirely. Some news reports even claimed—falsely—that Kansas had prohibited the teaching of evolution or mandated the teaching of biblical creationism.

In the national outcry that followed, Herbert Lin of the National Research Council (an affiliate of the National Academy of Sciences) wrote to *Science* suggesting that American colleges and universities should declare "their refusal to count as an academic subject any high school biology course taught in Kansas." The following month, *Scientific American* editor John Rennie recommended that college admissions committees tell Kansas school officials that "the qualifications of any students applying from that state in the future will have to be considered very carefully. Send a clear message to the parents in Kansas that this bad decision carries consequences for their children." Apparently, for Lin and Rennie, the need to enforce Darwinian orthodoxy justifies the academic equivalent of holding children hostage.

The truth is that a surprising number of biologists quietly doubt or reject some of the grander claims of Darwinian evolution. But—at least in America—they must keep their mouths shut or risk condemnation, marginalization, and eventual expulsion from the scientific community. This happens infrequently, but often enough to remind everyone that the risk is real. Even so, there is a growing underground of biologists who are disenchanted with the Darwinists' censorship of opposing viewpoints. When isolated dissidents begin to realize how many of their colleagues feel the same way, more and more of them will begin to speak out.

Ideally, biologists will then begin to clean their own house. Although the National Academy of Sciences has published booklets on evolution that blatantly misrepresent the truth, this does not mean that most of its members approve of concealing and distorting scientific evidence. It seems more likely that a relatively small faction in the National Academy—albeit with the approval of its current president, textbook-writer Bruce Alberts—has exploited the Academy's reputation to propagate Darwinian dogma. Once the distinguished scientists who make up the National Academy realize what is being done in their names, they will presumably take steps to correct the abuse.

But they might not. All Americans—including those in the National Academy of Sciences—are guaranteed the right to believe and speak as they choose. Scientists would be completely within their constitutional rights if they chose to continue supporting the present Darwinian establishment and its distortions of the truth. Unless they have your consent, however, they are not entitled to do it with your money.

It's your money

If you are a U.S. taxpayer, most of the financial support for the Darwinian establishment and its censorship of opposing viewpoints comes out of your pocket. The vast majority of research done by Darwinists in the United States is funded by agencies of the Federal Government, primarily the National Institutes of Health (NIH) and the National Science Foundation (NSF); and much of the funding for origin-of-life research comes from the National Aeronautics and Space Administration (NASA).

The year 2000 budget for the NIH was almost $18 billion; for the NSF, almost $4 billion; and for NASA, more than

$13 billion. Much of this $35 billion went to legitimate research on other issues, but a significant chunk of it went to research on Darwinian evolution. Unfortunately, it may be difficult for American taxpayers to determine exactly how much of their money is spent on such research. According to evolutionary biologist Douglas Futuyma, it has been "rumored that the National Science Foundation, sensitive to scrutiny by congressional watchdogs, has recommended that the word 'evolution' not be used in the titles of abstracts of grant applications."

Whether or not this rumor is true, there is no question that you are paying for most of the Darwinian research done in the United States. If you doubt this, simply pick up a biology journal at a university library, find some articles dealing with evolution, and turn to their acknowledgments. Most articles on evolution published by Americans acknowledge financial support from the NIH, NSF, or NASA.

Of course, research—even research on evolution—is not a bad thing. But as we saw in several of the icons of evolution, data are frequently claimed to support evolutionary theory even when they contradict it. If an article in a mainstream journal reports evidence inconsistent with Darwinian evolution, chances are that the authors explain it away and defend the orthodox position anyway—otherwise, their article might never have been published. And they're doing it with your money.

Tax dollars support not only journal articles, but also the teaching careers of the people who write them. The next time you see a recent issue of *Science,* pick it up and flip through the job ads in the back. Most applicants for college biology teaching jobs in the United States are expected to have (or be able to get) "extramural" or "external" funding in the form of research grants, most of which come from the U.S. government. Once the applicant is hired, the institution takes a thick slice of the pie to subsidize its own expenses. These are the schools where future biologists are being taught falsehoods and circular reasoning in the guise of science. Even if you don't have college-age children, your taxes are supporting these institutions and the dogmatic Darwinists who teach in them.

Federal support for research and teaching is not the only way you are compelled to support what amounts to a massive indoctrination campaign by dogmatic Darwinists. Through your state and local taxes, you are paying for a state university system, local community colleges, and public schools, all of which are teaching the icons of evolution as though they were facts. If you doubt this, go look at their textbooks. High school biology books generally cost over $40 apiece, because they include lots of full-color pictures. Now that you've read the truth about the icons of evolution, stop by your local high school sometime and see how your tax dollars are working for you.

If you are putting a son or daughter through college, some of your money may also be paying for college biology textbooks, most of which cost over $75 apiece. If those textbooks deal with evolution, you can bet that they contain at least some of the icons described in this book. When you add up federal and state tax support for research and teaching, state and local money spent on biology textbooks, and family support for students, you can see that the Darwinian establishment is receiving tens of billions of dollars annually from the American people.

What can you do about it?

If you object to supporting dogmatic Darwinists that misrepresent the truth to keep themselves in power, there may be things you can do about it. One possibility is to call for congressional hearings on the way federal money is distributed by the NIH, the NSF, and NASA. When Harvard biologist Louis Guenin wrote that "we describe breaches of candour as deception," he also wrote that "the government might reasonably assert that one who stoops to deception in quest of distinction betrays such lack of distinction that further support would waste public funds." Scientists who deliberately distort the evidence should be disqualified from receiving public funds.

As we have seen, the National Academy of Sciences publishes booklets that misrepresent the evidence for evolution. Although the National Academy is not a government agency, it receives about 85 percent of its funding from contracts with the federal government, and its finances are reviewed every year by the Judiciary Committee of the U.S. House of Representatives. Maybe your representatives should look more closely at how your money is being spent.

The U.S. Congress has already taken note of how dogmatic Darwinists treat dissenters in American academia. After the international conference on the conceptual foundations of science at Baylor University in April 2000, eight Baylor scientists (purporting to speak for the university as a whole) wrote to U.S. Representative Mark Souder (R-Indiana) to complain about the Michael Polanyi Center. Their letter backfired, however, when Souder blasted them on the floor of the U.S. House of Representatives. "As the Congress," Souder said, "it might be wise for us to question whether the legitimate authority of science over scientific matters is being misused by persons who wish to identify science with a philosophy they prefer. Does the scientific community really welcome new ideas and dissent, or does it merely pay lip service to them while imposing a materialist orthodoxy?"

State legislators might also want to take a look at the Darwinian establishment, to determine whether state taxes are being used for indoctrination rather than education. State and local school boards could be encouraged to take a closer look at the textbooks they buy for public schools. Textbooks already in circulation will probably continue to be used for a while—after all, it will be expensive to replace them, and most of the material in them is reasonably accurate anyway. But school boards might want to alert students to their misrepresentations by attaching warning labels.

Not all the financial support for dogmatic Darwinists is coerced from taxpayers. Voluntary donations by college graduates to their alma maters often go to departments that indoctrinate students in Darwinism rather than show them the real evidence. The next time you get a fundraising letter from your alma mater, you might want to ask where your money will go.

The danger with a popular revolt against the Darwinian establishment is that the baby might be thrown out with the bath. It's vitally important to remember that science is not the enemy. Publicly funded scientific research and high-quality science education are essential to the future well-being of our society. It would be a great tragedy if the excesses of dogmatic Darwinists provoked a public outcry that resulted in lowering support for scientific research in general. This is why biologists, most of whom are truth-seekers rather than dogmatists, will presumably want to take the lead in cleaning their own house.

Another reason for biologists to clean their own house is so they can avoid replacing one dogmatism with another. Some dogmatic Darwinists have been very effective at shoring up their monopoly by playing on the fear of religious fundamentalism. Darwinism is indispensable, we are told, because it protects us from religious fanatics who might impose a suffocating ortho-doxy on science. Ironically, these people "protect" science from religious dogmatism by imposing a dogmatism of their own. Nevertheless, it would be a shame if their dogmatism were sim-ply replaced by another.

So biologists will want to clean their own house before the taxpaying public has to do it for them, and they will want to avoid dogmatism altogether. The safest and best approach would simply be to restore biological science to its true foundation—the evidence.

Nothing in biology makes sense except in the light of WHAT?

In 1973, neo-Darwinist Theodosius Dobzhansky announced that "nothing in biology makes sense except in the light of evo-lution." Ever since, Dobzhansky's maxim has been the rallying cry for people who think that everything in biology should revolve around evolutionary theory.

Certainly, there are some areas of biology in which Darwin-ian evolution plays an important role. As we have seen, there is good evidence that mutations and natural selection are significant factors at the molecular level, especially in rendering bacteria resistant to antibiotics, or insects and other pests resistant to pesticides. There is also good evidence that natural selection can produce limited modifications within existing species such as Darwin's finches. Surely, anyone who wants to make sense of these phenomena would be foolish to ignore evolutionary theory.

Promoters of Darwinism typically use evidence from antibi-otic and pesticide resistance, and minor modifications within species, to justify their claim that the economically important fields of medicine and agriculture depend on their theory. Yet for most practical purposes Darwinian evolution is irrelevant to medicine—even in dealing with antibiotic resistance. A physician treating a patient with a bacterial infection usually begins by administering an antibiotic known to work in similar cases. If the antibiotic is ineffective, the physician may ask a laboratory technologist to identify the organism using biochemical tests, and determine what antibiotics would be more effective in combating it. But neither the physician nor the technologist needs evolutionary theory to diagnose or treat the infection.

Agriculture has also been quite successful without help from Darwinism. Of course, the domestic breeding of crops and livestock is important, but agricultural science was around long before Darwin. Even when it comes to pesticide resistance, farmers (like physicians) deal with problems pragmatically, on a case-by-case basis. Ironically, despite the Darwinists' insistence that nothing in agriculture makes sense without them, they were handed their greatest defeat in recent years by the State of Kansas—home of some of the most successful farmers in the world.

No one would deny that medicine and agriculture do best when they proceed scientifically. But science is not synonymous with Darwinism—contrary to what some dogmatic Darwinists would have us believe.

There are many other areas of biology which do quite well without Darwinian evolution. In fact, most major disciplines in modern biology—including embryology, anatomy, physiology, paleontology and genetics—were pioneered by scientists who had never heard of Darwinian evolution—or who (like von Baer) explicitly rejected it. Although Darwinian jargon has become commonplace in these fields in recent years, it is misleading and doctrinaire to say that nothing in them makes sense except in the light of evolution.

Evolutionary biologist Peter Grant (famous for his research on Darwin's finches) acknowledged in his presidential address to the American Society of Naturalists in 1999 that "not all biologists who would call themselves naturalists pay attention to [Dobzhansky's maxim] or even feel the need to. For example, an ecologist's world can make perfect sense, in the short term at least, in the absence of evolutionary considerations."

So the claim that "nothing in biology makes sense except in the light of evolution" is demonstrably false. A person can be a first-rate biologist without being a Darwinist. In fact, a person who rejects Dobzhansky's claim can be a better biologist than one who accepts it uncritically. The distinctive feature and greatest virtue of natural science, we are told, is its reliance on evidence. Someone who starts with a preconceived idea and distorts the evidence to fit it is doing the exact opposite of science. Yet this is precisely what Dobzhansky's maxim encourages people to do.

The icons of evolution are a logical consequence of the dogma that nothing in biology makes sense except in the light of evolution. All the misleading claims we have examined in this book follow from the sort of thinking represented by Dobzhansky's profoundly anti-scientific starting-point. The primitive atmosphere was strongly reducing. All organisms are descended from a universal common ancestor. Homology is similarity due to common ancestry, vertebrate embryos are most similar in their earliest stages, and birds are feathered dinosaurs. Peppered moths rest on tree trunks, natural selection produced fourteen species of Darwin's finches, mutations provide the raw materials for morphological evolution, and humans are accidental by-products of undirected natural processes.

How do we know all these things? Because of the evidence? No, because—Dobzhansky says—nothing in biology makes sense except in the light of evolution.

This is not science. This is not truth-seeking. This is dogmatism, and it should not be allowed to dominate scientific research and teaching. Instead of using the icons of evolution to indoctrinate students in Darwinian theory, we should be using them to teach students how theories can be corrected in light of the evidence. Instead of teaching science at its worst, we should be teaching science at its best.

And science at its best pursues the truth. Dobzhansky was dead wrong, and so are those who continue to chant his antiscientific mantra. To a true scientist, nothing in biology makes sense except in the light of evidence.

An Evaluation of Ten Recent Biology Textbooks

on their Use of Selected Icons of Evolution

(For textbooks and evaluation criteria, see the following pages.)

Textbook:	1	2	3	4	5	6	7	8	9	10
Icon:										
Miller–Urey experiment	D	D	F	F	D	F	D	F	F	F
Darwin's tree of life	F	D	D	F	F	F	F	F	F	F
Vertebrate limb homology	D	D	D	D	F	F	D	F	D	D
Haeckel's embryos	F	D	F	F	F	D	F	F	F	F
Archaeopteryx	C	B	D	D	D	F	D	F	F	F
Peppered moths	X	n/a	D	F	F	F	F	D	F	F
Darwin's finches	F	D	D	X	D	F	F	D	F	F
OVERALL RATING	D–	D+	D–	F	F	F	F	F	F	F

X = contains no image, but uncritically repeats the standard story in the text.

n/a = book contains no image or mention of this icon.

The overall rating is an average grade based on A = 4, B = 3, C = 2, D = 1, F = 0 and X = 1/2.

List of textbooks

(All have copyright dates of 1998 or later. Books are listed alphabetically by first author's last name.)

1. Alton Biggs, Chris Kapicka & Linda Lundgren, *Biology: The Dynamics of Life* (Westerville, OH: Glencoe/McGraw-Hill, 1998). ISBN 0-02-825431-7

2. Neil A. Campbell, Jane B. Reece & Lawrence G. Mitchell, *Biology*, Fifth Edition (Menlo Park, CA: The Benjamin/Cummings Publishing Company, 1999).
   ISBN 0-8053-6573-7

3. Douglas J. Futuyma, *Evolutionary Biology*, Third Edition (Sunderland, MA: Sinauer Associates, 1998).
   ISBN 0-87893-189-9

4. Burton S. Guttman, *Biology*, (Boston: WCB/McGraw-Hill, 1999).
   ISBN 0-697-22366-3

5. George B. Johnson, *Biology: Visualizing Life*, Annotated Teacher's Edition (Orlando, FL: Holt, Rinehart & Winston, 1998).
   ISBN 0-03-016724-8

6. Sylvia Mader, *Biology*, Sixth Edition (Boston: WCB/McGraw-Hill, 1998).
   ISBN 0-697-34080-5

7. Kenneth R. Miller & Joseph Levine, *Biology*, Fifth Edition (Upper Saddle River, NJ: Prentice-Hall, 2000).
   ISBN 0-13-436265-9

8. Peter H. Raven & George B. Johnson, *Biology*, Fifth Edition (Boston: WCB/McGraw-Hill, 1999).
   ISBN 0-697-35353-2

9. William D. Schraer & Herbert J. Stoltze, *Biology: The Study of Life*, Seventh Edition (Upper Saddle River, NJ: Prentice Hall, 1999).
   ISBN 0-13-435086-3

10. Cecie Starr & Ralph Taggart, *Biology: The Unity and Diversity of Life*, Eighth Edition (Belmont, CA: Wadsworth Publishing Company, 1998).
    ISBN 0-534-53001-X

Specific evaluation criteria

In general, an "A" requires full disclosure of the truth, discussion of relevant scientific controversies, and a recognition that Darwin's theory—like all scientific theories—might have to be revised or discarded if it doesn't fit the facts. An "F" indicates that the textbook uncritically relies on logical fallacy, dogmatically treats a theory as an unquestionable fact, or blatantly misrepresents published scientific evidence.

The Miller-Urey experiment

A = does not include a picture or drawing of the Miller-Urey apparatus, or else accompanies it with a caption pointing out that the experiment (though historically interesting) is probably irrelevant to the origin of life because it did not simulate conditions on the early Earth; text mentions the controversy over oxygen in the primitive atmosphere, and includes extensive discussion of the other problems faced by origin-of-life research, acknowledging that they remain intractable.

B = does not include a picture or drawing of the Miller-Urey apparatus, or else accompanies it with a caption pointing out that the experiment (though historically interesting) is probably irrelevant to the origin of life because it did not simulate conditions on the early Earth; text includes at least some discussion of other problems in origin-of-life research, and does not leave the student with the impression that scientists are on the verge of understanding the origin of life.

C = includes a picture or drawing of the Miller-Urey apparatus, but the caption does *not* claim that the Miller-Urey experiment simulated conditions on the early Earth; the accompanying text points out that the experiment fails even if other starting mixtures are used, and does *not* leave the student with the impression that the experiment (or some variant of it) demonstrated how life's building-blocks formed on the early earth; does not discuss other problems with origin-of-life research.

D = includes a picture or drawing of the Miller-Urey apparatus with a misleading caption claiming or implying that the experiment simulated conditions on the early Earth; but the accompanying text *explicitly* points out that this was probably not the case (merely listing other gasses, and leaving it to the student to spot the discrepancy, is not sufficient); may leave the student with the impression that the experiment (or some variant of it) demonstrated how life's building-blocks formed on the early earth.

F = includes a picture or drawing of the Miller-Urey apparatus with a misleading caption claiming or implying that the experiment simulated conditions on the early Earth; the text contains no mention of the experiment's flaws, and leaves the student with the impression that it demonstrated how life's building-blocks formed on the early earth.

Darwin's tree of life

A = explicitly treats universal common ancestry as a theory rather than a fact; clearly points out that the "top-down" Cambrian explosion contradicts the "bottom-up" pattern of Darwinian evolution, and acknowledges the theoretical possibility of multiple origins and separate lines of descent; also mentions problems for universal common ancestry posed by recent evidence from molecular phylogeny.

B = explicitly treats universal common ancestry as a theory rather than a fact; clearly points out that the "top-down" Cambrian explosion contradicts the "bottom-up" pattern of Darwinian evolution, and acknowledges the theoretical possibility of multiple origins and separate lines of descent; but does not mention recent problems in molecular phylogeny.

C = explicitly treats universal common ancestry as a theory rather than a fact; discusses the Cambrian explosion as a problem for Darwinian evolution, but does not mention the theoretical possibility of multiple origins and separate lines of descent.

D = assumes the truth of universal common ancestry without questioning it (and may call it a "fact"); mentions the Cambrian explosion in the body of the text (briefly mentioning it in a note at the end of the chapter, without explaining what it is, is not sufficient), but does not discuss the problem it poses for Darwinian evolution.

F = assumes the truth of universal common ancestry without questioning it (and may call it a "fact"); does not even mention the Cambrian explosion.

Homology in vertebrate limbs

A = defines homology as similarity of structure and position, and explains that this was historically attributed to a common archetype; mentions a biological ancestor as one possible meaning of "archetype," but acknowledges that there are others, and that the concept of homology continues to be controversial; clearly explains that the two biological mechanisms proposed so far to account for homology (similar genes and similar developmental pathways) are inconsistent with the evidence.

B = defines homology as similarity of structure and position due to a common archetype, and identifies "archetype" with a biological ancestor without explaining that there are other possibilities; points out that the two biological mechanisms proposed so far to account for it (similar genes and similar developmental pathways) are inconsistent with the evidence.

C = defines homology as similarity of structure and position, and cites it as evidence for common ancestry; attributes homology to similar genes or similar developmental pathways, but at least hints that there are problems with the evidence.

D = defines homology as similarity of structure and position, and cites it as evidence for common ancestry; may attribute homology to similar genes or similar developmental pathways, but fails to mention that the evidence does not fit the claim.

F = defines homology as similarity due to common ancestry, then engages in circular reasoning by citing homology as evidence for common ancestry.

Haeckel's embryos

A = does not use misleading drawings or photos, and does not call pharyngeal pouches "gill slits"; points out that vertebrate embryos are most similar midway through development, after being dissimilar in their earliest stages; acknowledges this as an unresolved problem for Darwinian evolution, and considers the possibility that Darwin's theory of vertebrate origins could be wrong.

B = does not use misleading drawings or photos, and does not call pharyngeal pouches "gill slits"; points out that vertebrate embryos are most similar midway through development, after being dissimilar in their earliest stages; acknowledges this as an unresolved problem for Darwinian evolution, but does not explicitly consider the possibility that Darwin's theory of vertebrate origins could be wrong.

C = does not use misleading drawings or photos; points out that vertebrate embryos are most similar midway through development, after being dissimilar in their earliest stages, but explains away this fact in order to reconcile it with Darwinian evolution; may call pharyngeal pouches "gill slits."

D = uses actual photos rather than Haeckel's drawings, but chooses those which best fit the theory; fails to mention that earlier stages are dissimilar, and claims that early similarities in vertebrate embryos are evidence for common ancestry and Darwinian evolution; may call pharyngeal pouches "gill slits."

F = uses Haeckel's drawings (or a re-drawn version of them) without mentioning the dissimilarity of earlier stages; claims that early similarities in vertebrate embryos are evidence for common ancestry and Darwinian evolution; may call pharyngeal pouches "gill slits."

Archaeopteryx: *the missing link*

A = explains that the status of *Archaeopteryx* as a transitional link between reptiles and birds is controversial; points out that modern birds are probably not descended from it; mentions the controversy over whether birds evolved from dinosaurs or from a more primitive group; points out that the supposed dinosaur ancestors of *Archaeopteryx* do not appear in the fossil record until tens of millions of years after it.

B = explains that the status of *Archaeopteryx* as a transitional link between reptiles and birds is controversial; points out that modern birds are probably not descended from it; mentions the controversy over whether birds evolved from dinosaurs or from a more primitive group; but fails to point out that the supposed dinosaur ancestors of *Archaeopteryx* do not appear in the fossil record until tens of millions of years after it.

C = explains that the status of *Archaeopteryx* as a transitional link between reptiles and birds is controversial; points out that modern birds are probably not descended from it; but does not mention the controversy over whether birds evolved from dinosaurs or from a more primitive group.

D = presents *Archaeopteryx* as the transitional link between reptiles (or dinosaurs) and modern birds; does not point out that modern birds are probably not descended from it, but at least hints at the fact that there is a controversy over its ancestry or its transitional status.

F = presents *Archaeopteryx* as the transitional link between reptiles (or dinosaurs) and modern birds; does not point out that modern birds are probably not descended from it, and does not even hint at the fact that there is a controversy over its ancestry or its transitional status.

Peppered moths

A = uses photos of moths in their natural resting places; does not use staged photos of moths on tree trunks (except as illustrations of how the classical story was wrong); clearly discusses unresolved problems with Kettlewell's experiments and the classical story, and points out that these problems raise serious doubts about whether peppered moths provide direct evidence for natural selection.

B = uses photos of moths in their natural resting places; does not use staged photos of moths on tree trunks (except as illustrations of how the classical story was wrong); mentions unresolved problems with Kettlewell's experiments and the classical story, but does not discuss the possibility that peppered moths do not provide direct evidence for natural selection.

C = uses staged photos but clearly explains that they were staged, because moths do not rest on tree trunks in the wild;

describes Kettlewell's experiments, but briefly mentions that they and the classical story are now in doubt.

D = uses staged photos without mentioning that they misrepresent the natural situation; but the accompanying text at least hints at the fact that there are problems with Kettlewell's experiments or the classical story.

F = uses staged photos without mentioning that they misrepresent the natural situation; describes Kettlewell's experiments as a demonstration of natural selection, without mentioning their flaws or problems with the classical story.

Darwin's finches

A = explicitly points out that the Galápagos finches had little to do with the formulation of Darwin's theory; explains that selection on finch beaks oscillates between wet and dry years, producing no net evolutionary change; points out *both* that the genes affecting finch beaks are unknown *and* that hybrids between several species are now more fit than their parents, suggesting that those species may be merging.

B = explicitly points out that the Galápagos finches had little to do with the formulation of Darwin's theory; explains that selection on finch beaks oscillates between wet and dry years, producing no net evolutionary change; points out *either* that the genes affecting finch beaks are unknown *or* that hybrids between several species are now more fit than their parents, suggesting that those species may be merging.

C = describes the Galápagos finches as a good example of adaptive radiation (the origin of species by natural selection); but points out *both* that selection on finch beaks oscillates between wet and dry years *and* that the finches did not play an important role in the formulation of Darwin's theory.

D = describes the Galápagos finches as a good example of adaptive radiation (the origin of species by natural selection); but points out *either* that selection on finch beaks oscillates between wet and dry years *or* that the finches did not play an important role in the formulation of Darwin's theory.

F = describes the Galápagos finches as a good example of adaptive radiation (the origin of species by natural selection); but fails to mention that selection on finch beaks oscillates between wet and dry years, and implies that the finches played an important role in the formulation of Darwin's theory.

Suggested Warning Labels for Biology Textbooks

Biology textbooks contain a wealth of valuable information. Just because they misrepresent the evidence for evolution doesn't mean that everything they teach is incorrect. Existing textbooks can and should be used until publishers come out with corrected ones. In the meantime, students should be warned, where necessary, that their books misrepresent the truth. Warning labels such as those below can be used for this purpose, but they should be applied only by, or under the direction of, the owner of the book.

> **WARNING:** The Miller–Urey experiment probably did not simulate the Earth's early atmosphere; it does not demonstrate how life's building-blocks originated.

> **WARNING:** Darwin's tree of life does not fit the fossil record of the Cambrian explosion, and molecular evidence does not support a simple branching-tree pattern.

WARNING: If homology is defined as similarity due to common ancestry, it cannot be used as evidence for common ancestry; whatever its cause may be, it is not similar genes.

WARNING: These pictures make vertebrate embryos look more similar than they really are; it is not true that vertebrate embryos are most similar in their earliest stages.

WARNING: *Archaeopteryx* is probably not the ancestor of modern birds, and its own ancestors remain highly controversial; other missing links are now being sought.

WARNING: Peppered moths do not rest on tree trunks in the wild, and photos showing them on tree trunks have been staged; Kettlewell's experiments are now being questioned.

WARNING: The Galápagos finches did not inspire Darwin with the idea of evolution, and oscillating natural selection on their beaks produces no observable net change.

WARNING: Four-winged fruit flies must be artificially bred, and their extra wings lack muscles; these disabled mutants are not raw materials for evolution.

WARNING: Evidence from fossil horses does not justify the claim that evolution was undirected, which is based on materialistic philosophy rather than empirical science.

WARNING: Theories about human origins are subjective and controversial, and they rest on little evidence; all drawings of "ancestors" are hypothetical.

Research Notes

Chapter 1: Introduction

The opening quotations are from Linus Pauling, *No More War!* (New York: Dod, Mead & Company, 1958), p. 209; Bruce Alberts, "Science and Human Needs," address delivered to the 137th Annual Meeting of the National Academy of Sciences, Washington, DC, May 1, 2000, which can be found online at www4.national academies.org/nas/na; Roger Lewin, *Bones of Contention*, Second Edition (Chicago: The University of Chicago Press, 1997), p. 318.

The discipline of science

The quotations from the National Academy of Sciences booklet on the nature of science are from *Teaching About Evolution and the Nature of Science* (Washington, DC: National Academy Press, 1998); the order in which they appear here is Chapter 5, p. 5; Chapter 4, p. 8; Chapter 3, p. 10. The booklet is available online at www.nap.edu/readingroom/books/evolution98.

The Bacon reference is a paraphrase of Francis Bacon, *Novum Organum, or True Suggestions for the Interpretation of Nature*, Aphorisms, Book I, p. 129.

The need for public scrutiny

The quotations by and about Jefferson are from the National Academy's *Teaching About Evolution and the Nature of Science* (1998), Preface, p. 1.

The Graham quotation is from U.S. District Judge James Graham, "Government Shouldn't Choose Sides in Evolution Debate," *The Columbus* [Ohio] *Dispatch* (May 13, 2000), p. 11A.

What is evolution?

Quotations from the National Academy's 1998 booklet that deal with the meaning of evolution are from Chapter 5, p. 1. Despite the prestigious auspices under which they wrote, the authors of the National Academy booklet employed the usual evasions in their defense of evolution. For example: "Evolution in its broadest sense explains that what we see today is different from what existed in the past." And: "There is no debate within the scientific community over whether evolution occurred, and there is no evidence that evolution has not occurred." (Chapter 5, p. 1; Chapter 1, p. 3)

Biology students are sometimes encouraged to respond to critics of Darwinian evolution by evading the issue: "When you hear someone wonder about whether 'evolution' takes place," wrote Cecie Starr and Ralph Taggart in their 1998 biology textbook, "remind yourself that evolution simply means *genetic change through time.* Selective breeding practices provide abundant, tangible evidence that heritable changes do, indeed, occur." From Cecie Starr and Ralph Taggart, *Biology: The Unity and Diversity of Life*, Eighth Edition (Belmont, CA: Wadsworth Publishing Company, 1998), p. 281. (emphasis in the original)

In this and subsequent chapters, all citations to Darwin's *Origin of Species* and *The Descent of Man* are from the Modern Library

Reprint Edition (New York: Random House, 1936). There were six editions of *The Origin of Species* from 1859 to 1872, and differences among them reveal interesting things about Darwin's intellectual journey, but (except for a brief reference in the chapter on Haeckel's embryos) this book does not deal with them. Since page numbers vary from edition to edition, all citations in this book list the chapter as well as page number. The quotations in this Introduction (in the order in which they appear) are from *The Origin of Species*, Conclusion (Chapter XV), p. 373; Introduction, p. 14.

Icons of evolution

The Stephen Jay Gould quotation is from *Wonderful Life* (New York: W. W. Norton, 1989), p. 28. This is part of the epigraph at the beginning of this book.

Chapter 2: The Miller–Urey Experiment

Charles Darwin's comment about the "warm little pond" was in an 1871 letter, reprinted in Francis Darwin (editor), *The Life and Letters of Charles Darwin* (New York: D. Appleton, 1887), Vol. 2, p. 202. See also A. I. Oparin, *Origin of Life* (Moscow, 1924; translated by S. Morgulis and published by Macmillan in 1938); J.B.S. Haldane, *Rationalist Annual* 148 (1928), pp. 3–10.

The Miller-Urey experiment

Harold Urey, "On the Early Chemical History of the Earth and the Origin of Life," *Proceedings of the National Academy of Sciences USA* 38 (1952), pp. 351–363; Stanley Miller, "A Production of Amino Acids Under Possible Primitive Earth Conditions," *Science* 117 (1953), pp. 528–529. See also Stanley Miller and Harold Urey,

"Organic Compound Synthesis on the Primitive Earth," *Science* 130 (1959), pp. 245–251.

Did the primitive atmosphere really lack oxygen?

On the source of the Earth's primitive atmosphere, see Harrison Brown, "Rare Gases and the Formation of the Earth's Atmosphere," pp. 258–266 in Gerard P. Kuiper (editor), *The Atmospheres of the Earth and Planets*, Revised Edition (Chicago: The University of Chicago Press, 1952); Heinrich D. Holland, "Model for the Evolution of the Earth's Atmosphere," pp. 447–477 in A. E. J. Engel, Harold L. James, and B. F. Leonard (editors), *Petrologic Studies: A Volume in Honor of A. F. Buddington* (Geological Society of America, 1962), pp. 448–449; Philip H. Abelson, "Chemical Events on the Primitive Earth," *Proceedings of the National Academy of Sciences USA* 55 (1966), pp. 1365–1372.

For arguments based on theoretical consequences of photodissociation, see L. V. Berkner and L. C. Marshall, "On the Origin and Rise of Oxygen Concentration in the Earth's Atmosphere," *Journal of the Atmospheric Sciences* 22 (1965), pp. 225–261; R. T. Brinkmann, "Dissociation of Water Vapor and Evolution of Oxygen in the Terrestrial Atmosphere," *Journal of Geophysical Research* 74 (1969), pp. 5355–5368; J. H. Carver, "Prebiotic atmospheric oxygen levels," *Nature* 292 (1981), pp. 136–138; James F. Kasting, "Earth's Early Atmosphere," *Science* 259 (1993), pp. 920–926.

On the uraninite evidence, see P. Ramdohr, "New observations on the ores of the Witwatersrand in South Africa and their genetic significance," *Transactions of the Geological Society of South Africa* (Annexure) 61 (1958), pp. 1–50; P. R. Simpson and J. F. W. Bowles, "Uranium mineralization of the Witwatersrand and Dominion Reef systems," *Philosophical Transactions of the Royal Society of London* A 286 (1977), pp. 527–548; D. E. Grandstaff, "Origin of the

Uraniferous Conglomerates at Elliot Lake, Canada and Witwatersrand, South Africa: Implications for Oxygen in the Precambrian Atmosphere," *Precambrian Research* 13 (1980), pp. 1–26. See also James C. G. Walker, Cornelis Klein, Manfred Schidlowski, J. William Schopf, David L. Stevenson, and Malcolm R. Walter, "Environmental Evolution of the Archean–Early Proterozoic Earth," pp. 260–290 in J. William Schopf (editor), *Earth's Earliest Biosphere* (Princeton: Princeton University Press, 1983). On the significance of iron deposits, see James C. G. Walker, *Evolution of the Atmosphere* (New York: Macmillan, 1977), p. 262; Erich Dimroth and Michael M. Kimberly, "Precambrian atmospheric oxygen: evidence in the sedimentary distributions of carbon, sulfur, uranium, and iron," *Canadian Journal of Earth Sciences* 13 (1976), pp. 1161–1185.

On the biochemical evidence, see J. Lumsden and D. O. Hall, "Superoxide dismutase in photosynthetic organisms provides an evolutionary hypothesis," *Nature* 257 (1975), pp. 670–672. See also Kenneth M. Towe, "Early Precambrian oxygen: a case against photosynthesis," *Nature* 274 (1978): 657–661; Robert M. Schwartz & Margaret O. Dayhoff, "Origins of Prokaryotes, Eukaryotes, Mitochondria, and Chloroplasts," *Science* 199 (1978), pp. 395–403. Controversy over the biochemical evidence has continued; see, for example, Jose Castresana and Matti Saraste, "Evolution of energetic metabolism: the respiration-early hypothesis," *Trends in Biochemical Sciences* 20 (1995), pp. 443–448.

Declaring the controversy over

For arguments that the Miller-Urey experiment justifies assuming that the primitive atmosphere lacked oxygen, see Sidney W. Fox and Klaus Dose, *Molecular Evolution and the Origin of Life,* Revised Edition (New York: Marcel Dekker, 1977), p. 44; James C. G. Walker,

Evolution of the Atmosphere (New York: Macmillan, 1977), p. 224; S. M. Awramik et al., "Biogeochemical Evolution of the Ocean-Atmosphere System State of the Art Report," pp. 309–320 in H. D. Holland and M. Schidlowski (editors), *Mineral Deposits and the Evolution of the Biosphere* (Berlin: Springer-Verlag, 1982), p. 310.

For the comment on "dogma" see Harry Clemmey and Nick Badham, "Oxygen in the Precambrian atmosphere: An evaluation of the geological evidence," *Geology* 10 (1982), pp. 141–146; Towe's comment is from Kenneth M. Towe, "Environmental Oxygen Conditions During the Origin and Early Evolution of Life," *Advances in Space Research* 18 (1996), pp. (12)7–(12)15.

For a review of the controversy over primitive oxygen levels, see Charles B. Thaxton, Walter L. Bradley, and Roger L. Olsen, *The Mystery of Life's Origin: Reassessing Current Theories* (Dallas, TX: Lewis and Stanley, 1984), Chapter 5.

The Miller-Urey experiment fails anyway

Heinrich D. Holland, "Model for the Evolution of the Earth's Atmosphere," pp. 447–477 in A. E. J. Engel, Harold L. James and B. F. Leonard (editors), *Petrologic Studies: A Volume in Honor of A. F. Buddington* (Geological Society of America, 1962), pp. 448–449; Philip H. Abelson, "Chemical Events on the Primitive Earth," *Proceedings of the National Academy of Sciences USA* 55 (1966), pp. 1365–1372.

For growing skepticism about a primitive reducing atmosphere, see Marcel Florkin, "Ideas and Experiments in the Field of Prebiological Chemical Evolution," *Comprehensive Biochemistry* 29B (1975), pp. 231–260; Sidney W. Fox and Klaus Dose, *Molecular Evolution and the Origin of Life,* Revised Edition (New York: Marcel Dekker, 1977), pp. 43, 74–76. See also James F. Kasting, "Earth's Early Atmosphere," *Science* 259 (1993), pp. 920–926; Jon Cohen,

"Novel Center Seeks to Add Spark to Origins of Life," *Science* 270 (1995), pp. 1925–1926; Heinrich D. Holland, *The Chemical Evolution of the Atmosphere and Oceans* (Princeton: Princeton University Press, 1984), pp. 99–100; Gordon Schlesinger and Stanley L. Miller, "Prebiotic Synthesis in Atmospheres Containing CH_4, CO, and CO_2: I. Amino Acids," *Journal of Molecular Evolution* 19 (1983), pp. 376–382; John Horgan, "In the Beginning...," *Scientific American* (February 1991), pp. 116-126; Joel Levine, "The Photochemistry of the Early Atmosphere," pp. 3-38 in Joel Levine (editor), *The Photochemistry of Atmospheres* (Orlando, FL: Academic Press, 1985), pp. 12–14.

An RNA world?

On the history of the RNA world hypothesis, see Kelly Kruger, Paula J. Grabowski, Arthur J. Zaug, Julie Sands, Daniel E. Gottschling, and Thomas R. Cech, "Self-Splicing RNA: Autoexcision and Autocyclization of the Ribosomal RNA Intervening Sequence of Tetrahymena," *Cell* 31 (1982), pp. 147–157; Cecilia Guerrier-Takada, Katheleen Gardiner, Terry Marsh, Norman Pace, and Sidney Altman, "The RNA Moiety of Ribonuclease P is the Catalytic Subunit of the Enzyme," *Cell* 35 (1983), pp. 849–857; Walter Gilbert, "The RNA world," *Nature* 319 (1986), p. 618.

On why RNA could not have been the first biomolecule, see Klaus Dose, "The Origin of Life: More Question than Answers," *Interdisciplinary Science Reviews* 13 (1988), pp. 348–356; Robert Shapiro, "Prebiotic Ribose Synthesis: A Critical Analysis," *Origins of Life and Evolution of the Biosphere* 18 (1988), pp. 71–85; Norman Pace, "Origin of Life—Facing Up to the Physical Setting," *Cell* 65 (1991), pp. 531–533; Leslie Orgel, "The origin of life—a review of facts and speculations," *Trends in Biochemical Sciences* (1998), pp. 491–495; Robert Shapiro, "Prebiotic cytosine synthesis: A critical analysis and implications for the origin of life," *Proceedings of the National Academy of Sciences USA* 96 (1999), pp. 4396–4401.

The Joyce quotations are from Gerald F. Joyce, "RNA evolution and the origins of life," *Nature* 338 (1989), pp. 217–224; Robert Irion, "RNA Can't Take the Heat," *Science* 279 (1998), p. 1303.

Quotations about the current state of origin-of-life research are from Klaus Dose, "The Origin of Life: More Questions Than Answers," *Interdisciplinary Science Reviews* 13 (1988), pp. 348–356; Leslie E. Orgel, "The origin of life: a review of facts and speculations," *Trends in Biochemical Sciences* 23 (1998), pp. 491–495; Nicholas Wade, "Life's Origins Get Murkier and Messier," *The New York Times* (Tuesday, June 13, 2000), pp. D1–D2. See also Gordon C. Mills, Malcolm Lancaster, and Walter L. Bradley, "Origin of Life and Evolution in Biology Textbooks—A Critique," *The American Biology Teacher* 55 (February, 1993), pp. 78–83.

The Miller-Urey experiment as an icon of evolution

Magazine and textbook quotations are from Richard Monastersky, "The Rise of Life on Earth," *National Geographic* 193 (March 1998), pp. 54–81; Kenneth R. Miller and Joseph Levine, *Biology*, Fifth Edition (Upper Saddle River, NJ: Prentice-Hall, 2000), pp. 343–344; William K. Purves, Gordon H. Orians, H. Craig Heller, and David Sadava, *Life: The Science of Biology*, Fifth Edition (Sunderland, MA: Sinauer Associates, 1998), Vol. 2, pp. 519–520; Douglas J. Futuyma, *Evolutionary Biology*, Third Edition (Sunderland, MA: Sinauer Associates, 1998), pp. 167, 169; Bruce Alberts et al., *Molecular Biology of the Cell*, Third Edition (New York: Garland Publishing, 1994), p. 4.

The quotation from the National Academy of Sciences booklet is in *Science and Creationism: A View from the National Academy of*

Sciences, Second Edition (Washington, DC: National Academy Press, 1999), Chapter on "The Origin of the Universe, Earth, and Life," p. 2.

The "mythology" criticism is from Robert Shapiro, *Origins: A Skeptic's Guide to the Creation of Life on Earth* (New York: Summit Books, 1986), p. 112.

Chapter 3: Darwin's Tree of Life

Darwin, *The Origin of Species*, Chapter XV, pp. 373, 370; Chapter IV, pp. 99–100; Chapter XV, pp. 370, 373; Figure 3–1 is from Chapter IV, p. 87. The Mayr quotation is from Ernst Mayr, *One Long Argument* (Cambridge, MA: Harvard University Press, 1991), p. 24.

Darwin's tree of life

Darwin, *The Origin of Species*, Chapter IV, pp. 99, 90, 91, 92–93; Chapter 15, p. 361.

Darwin and the fossil record

Darwin, *The Origin of Species*, Chapter X, pp. 252, 254, 255, 239.

The Cambrian explosion

On three billion year-old microfossils, see J. William Schopf, and Bonnie M. Packer, "Early Archean (3.3-Billion to 3.5-Billion-Year-Old) Microfossils from Warrawoona Group, Australia," *Science* 237 (1987), pp. 70–73.

On Ediacaran fossils, see Martin F. Glaessner, *The Dawn of Animal Life* (Cambridge: Cambridge University Press, 1984); Adolf Seilacher, "Late Precambrian and Early Cambrian Metazoa: preservational or real extinctions?" pp. 159–168 in H. D. Holland, and A. F. Trendall (editors), *Patterns of Change in Earth Evolution* (Berlin:

Springer Verlag, 1984); Gregory J. Retallak, "Were the Ediacaran fossils lichens?" *Paleobiology* 20 (1994), pp. 523–544; Stephen Jay Gould, *Wonderful Life* (New York: W. W. Norton, 1989), pp. 58–59; Simon Conway Morris, *The Crucible of Creation* (Oxford: Oxford University Press, 1998), p. 30.

On the interpretation of Ediacaran fossils, see also James W. Valentine and Douglas H. Erwin, "Interpreting Great Developmental Experiments: The Fossil Record," pp. 71–107 in Rudolf A. Raff and Elizabeth C. Raff (editors), *Development as an Evolutionary Process* (New York: Alan R. Liss, 1987). On the dating of the Ediacaran assemblage, see John P. Grotzinger, Samuel A. Bowring, Beverly Z. Saylor, and Alan J. Kaufman, "Biostratigraphic and Geochronologic Constraints on Early Animal Evolution," *Science* 270 (1995), pp. 598–604.

Stephen Jay Gould's *Wonderful Life* and Simon Conway Morris's *The Crucible of Creation* recount the story of the discovery and re-analysis of the Burgess Shale fossils. See also Harry B. Whittington, *The Burgess Shale* (New Haven, CT: Yale University Press, 1985); and Simon Conway Morris and H. B. Whittington, "The Animals of the Burgess Shale," *Scientific American* 241 (July, 1979), pp. 122–133. For some recent reports of Chengjiang fossils, see D-G. Shu, H-L. Luo, S. Conway Morris, X-L. Zhang, S-X. Hu, L. Chen, J. Han, M. Zhu, Y. Li, and L-Z. Chen, "Lower Cambrian vertebrates from south China," *Nature* 402 (1999), pp. 42–46; Jun-Yuan Chen, Di-Ying Huang and Chia-Wei Li, "An early Cambrian craniate-like chordate," *Nature* 402 (1999), pp. 518–522; Fred Heeren, "A Little Fish Challenges a Big Giant," *The Boston Globe* (May 30, 2000), p. E1.

For the dating of the Cambrian explosion, see Samuel A. Bowring, John P. Grotzinger, Clark E. Isachsen, Andrew H. Knoll, Shane M. Pelechaty, and Peter Kolosov, "Calibrating Rates of Early

Cambrian Evolution," *Science* 261 (1993), pp. 1293–1298. A more recent estimate puts the beginning of the Cambrian at 543 million years ago, thus shortening the Cambrian explosion by a million years; John P. Grotzinger, "The Terminal Proterozoic Time Scale: Constraints on Global Correlations and Rates of Early Animal Evolution," *American Association of Petroleum Geologists Bulletin* 81 (1997), pp. 1954.

Quotations are from James W. Valentine, Stanley M. Awramik, Philip W. Signor, and Peter M. Sadler, "The Biological Explosion at the Precambrian-Cambrian Boundary," *Evolutionary Biology* 25 (1991), pp. 279–356, pp. 279, 281. For additional accounts of the Cambrian explosion, see Mark A. S. McMenamin and Dianna L. Schulte McMenamin, *The Emergence of Animals: The Cambrian Breakthrough* (New York: Columbia University Press, 1990); Jere H. Lipps and Philip W. Signor (editors), *Origin and Early Evolution of the Metazoa* (New York: Plenum Press, 1992); Jeffrey S. Levinton, "The Big Bang of Animal Evolution," *Scientific American* 267 (November 1992), pp. 84–91; J. Madeleine Nash, "When Life Exploded," *Time* (December 4, 1995), pp. 66–74.

Figure 3–4: The names of the major living animal phyla, listed in alphabetical order, are:

(a) Annelida (clamworms, earthworms, leeches)
(b) Arthropoda (insects, crabs, centipedes, spiders)
(c) Brachiopoda (lamp shells)
(d) Bryozoa (small aquatic animals with tentacle-ringed mouths)
(e) Chaetognatha (arrow worms)
(f) Chordata (tunicates, lancelets, vertebrates)
(g) Cnidaria (corals, jellyfish, hydras)
(h) Ctenophora (comb jellies, sea walnuts)
(i) Echinodermata (crinoids, sea urchins, starfish, sea cucumbers)

(j) Hemichordata (acorn worms)

(k) Mollusca (clams, octopuses, snails)

(l) Nematoda (eelworms, roundworms)

(m) Onychophora (small terrestrial worms with short legs)

(n) Phoronida (tube-dwelling marine worms with tentacles)

(o) Platyhelminthes (flatworms, flukes, tapeworms)

(p) Pogonophora (giant deep-sea tube worms)

(q) Porifera (sponges)

(r) Rotifera (small animals with a crown of cilia)

Sponges (q) first appeared in the late Precambrian, and some paleontologists believe that Cnidaria (g) and Mollusca (k) did, too. Bryozoa (d) are first found in the Ordovician. All phyla shown as appearing in the Cambrian occur in the Lower Cambrian except Chaetognatha (e) and Hemichordata (j), which first appeared in the Middle Cambrian.

The challenge to Darwin's theory

Quotation about the "head of Zeus" is from Jeffrey H. Schwartz, "Homeobox Genes, Fossils, and the Origin of Species," *Anatomical Record (New Anatomist)* 257 (1999), pp. 15–31. See also Robert L. Carroll, "Towards a new evolutionary synthesis," *Trends in Ecology and Evolution* 15 (2000), pp. 27–32.

On top-down evolution, see Valentine, et al., *Evolutionary Biology* 25 (1991), pp. 279–356. On "phylogenetic lawns," see Conway Morris, *The Crucible of Creation,* p. 176.

Saving Darwin's theory

James W. Valentine and Douglas H. Erwin, "Interpreting Great Developmental Experiments: The Fossil Record," pp. 71–107 in Rudolf A. Raff and Elizabeth C. Raff (editors), *Development as an Evolutionary Process* (New York: Alan R. Liss, 1987), p. 84–85; Valentine, et al., *Evo-*

lutionary Biology 25 (1991), pp. 279–356, M. J. Benton, M. A. Wills, and R. Hitchin, "Quality of the fossil record through time," *Nature* 403 (2000), pp. 534–536. See also Mark A. Norell and Michael J. Novacek, "The Fossil Record and Evolution: Comparing Cladistic and Paleontologic Evidence for Vertebrate History," *Science* 255 (1992), pp. 1690–1693. On three billion year-old microfossils, see Andrew H. Knoll and Elso S. Barghoorn, "Archean Microfossils Showing Cell Division from the Swaziland System of South Africa," *Science* 198 (1977), pp. 396–398; J. William Schopf and Bonnie M. Packer, "Early Archean (3.3-Billion to 3.5-Billion-Year-Old) Microfossils from Warrawoona Group, Australia," *Science* 237 (1987), pp. 70–73.

Quotations are from Conway Morris, *The Crucible of Creation,* pp. 2, 28.

J. William Schopf, "The early evolution of life: solution to Darwin's dilemma," *Trends in Ecology and Evolution* 9 (1994), pp. 375–377. See also Valentine, et al., *Evolutionary Biology* 25 (1991), pp. 279–356; and Stefan Bengston, "The advent of animal skeletons," pp. 412–425 in Stefan Bengston (editor), *Early Life on Earth* (New York: Columbia University Press, 1994).

Molecular phylogeny

Emile Zuckerkandl and Linus Pauling, "Molecular Disease, Evolution, and Genetic Heterogeneity," pp. 189–225 in Michael Kasha and Bernard Pullman (editors), *Horizons in Biochemistry* (New York: Academic Press, 1962), pp. 200–201. See also Emile Zuckerkandl and Linus Pauling, "Molecules as Documents of Evolutionary History," *Journal of Theoretical Biology* 8 (1965), pp. 357–366; Emile Zuckerkandl and Linus Pauling, "Evolutionary Divergence and Convergence in Proteins," pp. 97–166 in Vernon Bryson and Henry J. Vogel (editors), *Evolving Genes and Proteins* (New York: Academic Press, 1965). Michael T. Ghiselin, "Models in Phylogeny," pp. 130–145 in Thomas J. M. Schopf (editor), *Models in Paleobiology* (San Francisco: Freeman, Cooper and Company, 1972), p. 145.

On methodological problems with molecular sequence analyses, see David P. Mindell, "Aligning DNA Sequences: Homology and Phylogenetic Weighting," pp. 73–89 in Michael M. Miyamoto and Joel Cracraft (editors), *Phylogenetic Analysis of DNA Sequences* (New York: Oxford University Press, 1991); John Gatesby, Rob DeSalle, and Ward Wheeler, "Alignment-Ambiguous Nucleotide Sites and the Exclusion of Systematic Data," *Molecular Phylogenetics and Evolution* 2 (1993), pp. 152–157; David M. Hillis, John P. Huelsenbeck, and Clifford W. Cunningham, "Application and Accuracy of Molecular Phylogenies," *Science* 264 (1994), pp. 671–677; Roderic Guigo, Ilya Muchnik, and Temple F. Smith, "Reconstruction of Ancient Molecular Phylogeny," *Molecular Phylogenetics and Evolution* 6 (1996), pp. 189–213; Arcady R. Mushegian, James R. Garey, Jason Martin, and Leo X. Liu, "Large-Scale Taxonomic Profiling of Eukaryotic Model Organisms," *Genome Research* 8 (1998), pp. 590–598; Laura E. Maley and Charles R. Marshall, "The Coming of Age of Molecular Systematics," *Science* 279 (1998), pp. 505–506.

Molecular phylogeny and the Cambrian explosion

On estimates of the date for the divergence of the animal phyla, see Bruce Runnegar, "A molecular clock date for the origin of the animal phyla," *Lethaia* 15 (1982), pp. 199–205; Russell F. Doolittle, Da-Fei Feng, Simon Tsang, Glen Cho, and Elizabeth Little, "Determining Divergence Times of the Major Kingdoms of Living Organisms with a Protein Clock," *Science* 271 (1996), pp. 470–477; Gregory A. Wray, Jeffrey S. Levinton and Leo H. Shapiro, "Molecular Evidence for Deep Precambrian Divergences Among Metazoan Phyla," *Science* 274 (1996), pp. 568–573; Richard A. Fortey, Derek E. G. Briggs, and Matthew A. Wills, "The Cambrian evolutionary 'explosion' recalibrated," *BioEssays* 19 (1997), pp. 429–434; Francisco José Ayala, Andrey Rzhetsky, and Francisco J. Ayala, "Origin of the metazoan phyla: Molecular clocks confirm paleontological estimates," *Proceedings of the National Academy of Sciences USA* 95 (1998), pp. 606–611; Kenneth M. Halanych, "Considerations for Reconstructing Metazoan History: Signal, Resolution, and Hypothesis Testing," *American Zoologist* 38 (1998), pp. 929–941. For a recent review, see Simon Conway Morris, "Evolution: Bringing Molecules into the Fold," *Cell* 100 (2000), pp. 1–11.

Lindell Bromham, Andrew Rambault, Richard Fortey, Alan Cooper, and David Penny, "Testing the Cambrian explosion hypothesis by using a molecular dating technique," *Proceedings of the National Academy of Sciences USA* 95 (1998), pp. 12386–12389. See also Andrew B. Smith, "Dating the origin of metazoan body plans," *Evolution and Development* 1 (1999), pp. 138–142; Simon Conway Morris, "Early Metazoan Evolution: Reconciling Paleontology and Molecular Biology," *American Zoologist* 38 (1998), pp. 867–877 (I have changed Conway Morris's "artefact" to "artifact" to conform to the American spelling and sharpen the contrast with Smith's statement, above).

James W. Valentine, David Jablonski, and Douglas H. Erwin, "Fossils, molecules and embryos: new perspectives on the Cambrian explosion," *Development* 126 (1999), pp. 851–859. See also Valentine et al., *Evolutionary Biology* 25 (1991), pp. 279–356; Simon Conway Morris, "The Cambrian 'explosion': Slow-fuse or megatonnage?" *Proceedings of the National Academy of Sciences USA* 97 (2000), pp. 4426–4429.

The growing problem in molecular phylogeny

For the classical view, see Carl R. Woese, "Bacterial Evolution," *Microbiological Reviews* 51 (1987), pp. 221–271.

Quotations about general problems are from James A. Lake, Ravi Jain, and Maria C. Rivera, "Mix and Match in the Tree of Life," *Science* 283 (1999), pp. 2027–2028, p. 2027; Hervé Philippe and Patrick Forterre, "The Rooting of the Universal Tree of Life Is Not Reliable," *Journal of Molecular Evolution* 49 (1999), pp. 509–523, p. 510; Carl Woese, "The universal ancestor," *Proceedings of the National Academy of Sciences USA* 95 (1998), pp. 6854–6859, p. 6854; Michael Lynch, "The Age and Relationships of the Major Animal Phyla," *Evolution* 53 (1999), pp. 319–325, p. 323.

On problems with mammal phylogenies, see Dan Graur, Laurent Duret, and Manolo Gouy, "Phylogenetic position of the order Lagomorpha (rabbits, hares and allies)," *Nature* 379 (1996), pp. 333–335; Gavin J. P. Naylor and Wesley M. Brown, "Amphioxus Mitochondrial DNA, Chordate Phylogeny, and the Limits of Inference Based on Comparisons of Sequences," *Systematic Biology* 47 (1998), pp. 61–76; Ying Cao, Axel Janke, Peter J. Waddell, Michael Westerman, Osamu Takenaka, Shigenori Murata, Norihiro Okada, Svante Pääbo, and Masami Hasegawa, "Conflict Among Individual Mitochondrial Proteins in Resolving the Phylogeny of Eutherian Orders," *Journal of Molecular Evolution* 47(1998), pp. 307–322. See also Michael P. Cummings, Sarah P. Otto, and John Wakeley, "Genes and Other Samples of DNA Sequence Data for Phylogenetic Inference," *Biological Bulletin* 196 (1999), pp. 345–350.

Uprooting the tree of life

Hervé Philippe and Patrick Forterre, "The Rooting of the Universal Tree of Life Is Not Reliable," *Journal of Molecular Evolution* 49 (1999), pp. 509–523, p. 520; Hervé Philippe and André Adoutte, "The molecular phylogeny of Eukaryota: solid facts and uncertainties," pp. 25–56 in G. H. Coombs, et al. (editors), *Evolutionary Relationships Among Protozoa* (Dordrecht, Netherlands: Kluwer Academic Publishers, 1998); Hervé Philippe and Jacqueline Laurent, "How good are deep phylogenetic trees?" *Current Opinion in Genetics and Development* 8 (1998), pp. 616–623; Patrick Forterre and Hervé Philippe, "Where is the root of the universal tree of life," *BioEssays* 21 (1999), pp. 871–879; Patrick Forterre and Hervé Philippe, "The Last Universal Common Ancestor (LUCA), Simple or Complex?" *Biological Bulletin* 196 (1999), pp. 373–377. See also Sarah A. Teichmann and Graeme Mitchison, "Is There a Phylogenetic Signal in Prokaryote Proteins?" *Journal of Molecular Evolution* 49 (1999), pp. 98–107; Laura A. Katz, "The Tangled Web: Gene Genealogies and the Origin of Eukaryotes," *The American Naturalist* 154 Supplement (1999), pp. S137–S145; Andrew J. Roger, "Reconstructing Early Events in Eukaryotic Evolution," *The American Naturalist* 154 Supplement (1999), pp. S146–S163.

Carl Woese, "The universal ancestor," *Proceedings of the National Academy of Sciences USA* 95 (1998), pp. 6854–6859; quotations (listed in the order in which they appear in the text) are from pp. 6854, 6855, 6858, 6856, 6854. See also Carl Woese, "Default taxonomy: Ernst Mayr's view of the microbial world," *Proceedings of the National Academy of Sciences USA* 95 (1998), pp. 11043–11046.

W. Ford Doolittle, "Phylogenetic Classification and the Universal Tree," *Science* 284 (1999), pp. 2124–2128; W. Ford Doolittle, "Uprooting the Tree of Life," *Scientific American* 282 (February, 2000), pp. 90–95. See also W. Ford Doolittle, "Lateral Genomics," *Trends in Biochemical Sciences* 24 (1999), M5–M8.

The fact of evolution?

For a photograph of the Hard Facts Wall see *Teaching Science in a Climate of Controversy* (Ipswich, MA: American Scientific Affiliation, 1986), p. 61. The photograph is accompanied by a description and explanatory drawings, pp. 56–63.

National Academy of Sciences, *Teaching About Evolution and the Nature of Science* (Washington, DC: National Academy Press, 1998),

Chapter 5, pp. 2–3; Chapter 2, pp. 5–6. National Academy of Sciences, *Science and Creationism: A View from the National Academy of Sciences*, Second Edition (Washington, DC: National Academy Press, 1999), "Evidence Supporting Biological Evolution," pp. 3, 6–7; J. Madeleine Nash, "When Life Exploded," *Time* (December 4, 1995), pp. 66–74.

Textbook quotations are from Douglas Futuyma, *Evolutionary Biology*, Third Edition (Sunderland, MA: Sinauer Associates, 1998), p. 15; Neil A. Campbell, Jane B. Reece, and Lawrence G. Mitchell, *Biology*, Fifth Edition (Menlo Park, CA: The Benjamin/Cummings Publishing Company, 1999), pp. 419, 426.

Quotations of dissenters from the "fact" of universal common descent are from Harry Whittington, *The Burgess Shale*, p. 131; Malcolm S. Gordon, "The Concept of Monophyly: A Speculative Essay," *Biology and Philosophy* 14 (1999), pp. 331–348.

The Chinese paleontologist story has been making the rounds since I first told it to some colleagues in 1999. Sadly, the principal reaction from dogmatic American Darwinists has been to demand his name. I refuse to give it to them, knowing what their colleagues have been doing to critics since at least 1981, when British paleontologist Colin Patterson, in a famous lecture at the American Museum of Natural History in New York, openly questioned whether there is any evidence for evolution. Afterwards, dogmatic Darwinists hounded him relentlessly, and Patterson never again voiced his skepticism in public. I fear they would do the same to the Chinese paleontologist in my story, an excellent scientist who deserves to be protected from heresy-hunters.

Chapter 4: Homology in Vertebrate Limbs

The terms "analogy" and "homology" did not actually originate with Owen, but with William MacLeay twenty years earlier (and the concepts themselves were much older). For the history of the concept, see Alec L. Panchen, "Richard Owen and the Concept of Homology," pp. 21–62 in Brian K. Hall (editor), *Homology: The Hierarchical Basis of Comparative Biology* (San Diego, CA: Academic Press, 1994). See also Peter J. Bowler, *Evolution: The History of an Idea,* Revised Edition (Berkeley: University of California Press, 1989). Quotations are from Charles Darwin, *The Origin of Species*, Chapter XIV, p. 335; Chapter XV, pp. 366, 352.

Re-defining homology

Ernst Mayr, *The Growth of Biological Thought* (Cambridge, MA: Harvard University Press, 1982), pp. 232, 465.

Homology and circular reasoning

J. H. Woodger, "On Biological Transformations," pp. 95–120 in W. E. Le Gros Clark and P. B. Medawar (editors), *Essays on Growth and Form Presented to D'Arcy Wentworth Thompson* (Oxford: Clarendon Press, 1945), p. 109; Alan Boyden, "Homology and Analogy," *American Midland Naturalist* 37 (1947), pp. 648–669. George Gaylord Simpson's procedures are in his *Principles of Animal Taxonomy* (New York: Columbia University Press, 1961); Robert R. Sokal and Peter H. A. Sneath criticize him for circularity in their *Principles of Numerical Taxonomy* (San Francisco: Freeman, 1963), p. 21.

Michael T. Ghiselin, "An Application of the Theory of Definitions to Systematic Principles," *Systematic Zoology* 15 (1966), pp. 127–130; Michael T. Ghiselin, "Models in Phylogeny," pp. 130–145 in Thomas J. M. Schopf (editor), *Models in Paleobiology* (San Francisco: Freeman, Cooper and Company, 1972), p. 134; David L. Hull, "Certainty and Circularity in Evolutionary Taxonomy," *Evolution* 21 (1967), pp. 174–189; N. Jardine, "The concept of homology in biology," *British Journal for the Philosophy of Science* 18 (1967), pp. 125–139; Donald H. Colless, "The Phylogenetic Fallacy," *Systematic Zoology* 16 (1967), pp. 289–295.

The "method of successive approximation," or "method of reciprocal illumination," was proposed by Willi Hennig, founder of cladistics; for a review of the "groping" criticism, see Sokal and Sneath, *Principles of Numerical Taxonomy*, p. 21. Biologist Walter Bock attempted to solve the problem in 1974 by embracing the neo-Darwinian definition of homology as common ancestry but specifying that "similarity between features is the only method for recognizing homologues." Walter J. Bock, "Philosophical Foundations of Classical Evolutionary Classification," *Systematic Zoology* 22 (1974), pp. 375–392. But this merely equivocated between two meanings of homology. On the logical problem in separating "definition" from "recognition criteria," see Bruce A, Young, "On the Necessity of an Archetypal Concept in Morphology: With Special Reference to the Concepts of 'Structure' and 'Homology'," *Biology and Philosophy* 8 (1993), pp. 225–248. Ronald H. Brady, "On the Independence of Systematics," *Cladistics* 1 (1985), pp. 113–126.

Breaking the circle

David B. Wake, "Homoplasy, homology and the problem of 'sameness' in biology," pp. 24–33 and 44–45 in *Homology* (Novartis Symposium 222; Chichester, UK: John Wiley & Sons, 1999), pp. 45, 27.

Evidence from DNA sequences

David M. Hillis, "Homology in Molecular Biology," pp. 339–368 in Brian K. Hall (editor), *Homology: The Hierarchical Basis of Comparative Biology*, pp. 339–341, 359. Colin Patterson, David M. Williams, and Christopher J. Humphries, "Congruence Between Molecular and Morphological Phylogenies," *Annual*

Review of Ecology and Systematics 24 (1993), pp. 153–188. See also Colin Patterson, "Homology in Classical and Molecular Biology," *Molecular Biology and Evolution* 5 (1988), pp. 603–625; and Michael S. Y. Lee, "Molecular phylogenies become functional," *Trends in Ecology and Evolution* 14 (1999), pp. 177–178. On the growing problems with DNA sequence comparisons, see the previous chapter and the recent article by W. Ford Doolittle, "Uprooting the Tree of Life," *Scientific American* 282 (February, 2000), pp. 90–95.

The fossil record

Sokal and Sneath, *Principles of Numerical Taxonomy*, pp. 56–57; Bruce A, Young, "On the Necessity of an Archetypal Concept in Morphology: With Special Reference to the Concepts of 'Structure' and 'Homology'," *Biology and Philosophy* 8 (1993), pp. 225–248, p. 231. See also Peter H. A. Sneath and Robert R. Sokal, *Numerical Taxonomy* (San Francisco, CA: W. H. Freeman and Company, 1973), p. 76; Elliot Sober, *Reconstructing the Past* (Cambridge, MA: MIT Press, 1988), p. 20.

For the Corvette analogy, see Tim Berra, *Evolution and the Myth of Creationism* (Stanford, CA: Stanford University Press, 1990), pp. 117–119. Berra's analogy was anticipated at least as early as 1964, but at that time it was used to illustrate the *problems* involved in inferring evolutionary relationships from similarities! See Rolf Sattler, "Methodological Problems in Taxonomy," *Systematic Zoology* 13 (1964), pp. 19–27. For a recent textbook use of Berra's Blunder, see T. Douglas Price and Gary M. Feinman, *Images of the Past*, Second Edition (Mountain View, CA: Mayfield Publishing, 1997), p. 3.

Phillip E. Johnson, *Defeating Darwinism by Opening Minds* (Downers Grove, IL: Intervarsity Press, 1997), pp. 62–63; Darwin, *The Origin of Species,* Introduction, p. 12; Chapter XIV, p. 343.

Leigh M. Van Valen, "Homology and Causes," *Journal of Morphology* 173 (1982), pp. 305–312.

Evidence from developmental pathways

Edmund B. Wilson, "The Embryological Criterion of Homology," pp.101–124 in *Biological Lectures Delivered at the Marine Biological Laboratory of Wood's Hole in the Summer Session of 1894* (Boston: Ginn & Company, 1895), p. 107; Gavin de Beer, *Embryos and Ancestors,* Third Edition (Oxford: Clarendon Press, 1958), p. 152. Pere Alberch, "Problems with the Interpretation of Developmental Sequences," *Systematic Zoology* 34 (1985), pp. 46–58; Rudolf Raff, "Larval homologies and radical evolutionary changes in early development," pp. 110–121 in *Homology* (Novartis Symposium 222; Chichester, UK: John Wiley & Sons, 1999), p. 111.

On the differences between salamanders and other vertebrates, see Neil H. Shubin and Pere Alberch, "A Morphogenetic Approach to the Origin and Basic Organization of the Tetrapod Limb," *Evolutionary Biology* 20 (1986), pp. 319–387; and Neil H. Shubin, "History, Ontogeny, and Evolution of the Archetype," pp. 249–271 in Brian K. Hall (editor), *Homology: The Hierarchical Basis of Comparative Biology*, pp. 264–266.

The Hinchliffe and Griffiths quotation is from J. R. Hinchliffe and P. J. Griffiths, "The prechondrogenic patterns in tetrapod limb development and their phylogenetic significance," pp. 99–121 in B. C. Goodwin, N. Holder and C. C. Wylie (editors), *Development and Evolution* (Cambridge: Cambridge University Press, 1983), p. 118. See also J. R. Hinchliffe, "Reconstructing the Archetype: Innovation and Conservatism in the Evolution and Development of the Pentadactyl Limb," pp. 171–189 in D. B. Wake and G. Roth (editors), *Complex Organismal Functions: Integration and Evolution in Vertebrates* (Chichester, UK: John Wiley & Sons, 1989); Neil H. Shubin,

"The Implications of 'The Bauplan' for Development and Evolution of the Tetrapod Limb," pp. 411–421 in J. R. Hinchliffe, J. M. Hurle, and D. Summerbell (editors), *Developmental Patterning of the Vertebrate Limb* (New York: Plenum Press, 1991).

The idea of an archetypal (ancestral) pattern common to all vertebrate limbs was proposed by Nils Holmgren, "On the Origin of the Tetrapod Limb," *Acta Zoologica* 14 (1933), pp. 185–295. Shubin and Hinchliffe suggest that it is developmental *processes*, rather than patterns, that are homologous. Brian Hall, however, argues that "homology is a statement about pattern, and should not be conflated with a concept about processes and mechanisms;" see Brian K. Hall, *Evolutionary Developmental Biology* (London: Chapman & Hall, 1992), p. 194. Günter Wagner maintains that the relevant processes are those that maintain patterns rather than generate them; see G. P. Wagner and B. Y. Misof, "How can a character be developmentally constrained despite variation in developmental pathways," *Journal of Evolutionary Biology* 6 (1993), pp. 449–455. The form-maintaining mechanisms, however, have yet to be identified. For a general review of the lack of correlation between homology and developmental pathways see Rudolf Raff, *The Shape of Life* (Chicago: The University of Chicago Press, 1996).

Evidence from developmental genetics

For de Beer's comment about homology and genes, see Gavin de Beer, *Homology: An Unsolved Problem* (London: Oxford University Press, 1971), pp. 15–16. On homologous features not due to homologous genes, see Gregory A. Wray and Ehab Abouheif, "When is homology not homology?" *Current Opinion in Genetics & Development* 8 (1998), pp. 675–680.

On developmental genes from mice functionally replacing their counterparts in flies, see Georg Halder, Patrick Callaerts, and

Walter J. Gehring, "Induction of Ectopic Eyes by Targeted Expression of the *eyeless* Gene in *Drosophila*," *Science* 267 (1995), pp. 1788–1792. See also Jarema Malicki, Klaus Shughart, and William McGinnis, "Mouse *Hox-2.2* Specifies Thoracic Segmental Identity in Drosophila Embryos and Larvae," *Cell* 63 (1990), pp. 961–967; Nadine McGinnis, Michael A. Kuziora, and William McGinnis, "Human *Hox-4.2* and Drosophila *Deformed* Encode Similar Regulatory Specificities in Drosophila Embryos and Larvae," *Cell* 63 (1990), pp. 969–976; Jack Jiagang Zhao, Robert A. Lazzarini, and Leslie Pick, "The mouse *Hox-1.3* gene is functionally equivalent to the *Drosophila Sex combs reduced* gene," *Genes & Development* 7 (1993), pp. 343–354. On *Distal-less*, see Grace Panganiban, et al., "The origin and evolution of animal appendages," *Proceedings of the National Academy of Sciences USA* 94 (1997), pp. 5162–5166; Gregory Wray, "Evolutionary dissociations between homologous genes and homologous structures," pp. 189–203 in *Homology* (Novartis Symposium 222; Chichester, UK: John Wiley & Sons, 1999), pp. 195–196.

On similar genetic networks, see Neil Shubin, Cliff Tabin, and Sean Carroll, "Fossils, genes and the evolution of animal limbs," *Nature* 388 (1997), pp. 639–648; and Clifford J. Tabin, Sean B. Carroll, and Grace Panganiban, "Out on a Limb: Parallels in Vertebrate and Invertebrate Limb Patterning and the Origin of Appendages," *American Zoologist* 39 (1999), pp. 650–663. See also Concepción Rodriguez-Esteban et al., "*Radical fringe* positions the apical ectodermal ridge at the dorsoventral boundary of the vertebrate limb," *Nature* 386 (1997), pp. 360–366; and Ed Laufer et al., "Expression of *Radical fringe* in limb-bud ectoderm regulates apical ectodermal ridge formation," *Nature* 386 (1997), pp. 366–373.

Gavin de Beer, *Homology: An Unsolved Problem* (London: Oxford University Press, 1971), p. 16.

Vertebrate limbs as evidence for evolution?

Teresa Audesirk and Gerald Audesirk, *Biology: Life on Earth* (Upper Saddle River, NJ: Prentice Hall, 1999), p. 264; Sylvia S. Mader, *Biology*, Sixth Edition (Boston: WCB/McGraw-Hill, 1998), p. 298; Peter H. Raven and George B. Johnson, *Biology*, Fifth Edition (Boston: WCB/McGraw-Hill, 1999), pp. 412, 416; Neil A. Campbell, Jane B. Reece, and Lawrence G. Mitchell, *Biology*, Fifth Edition (Menlo Park, CA: Addison Wesley Longman, 1999), p. 424. See also Helena Curtis and N. Sue Barnes, *Invitation to Biology*, Fifth Edition (New York: Worth Publishers, 1994), pp. 404–412.

Interestingly, Miller and Levine's *Biology* (Upper Saddle River, NJ: Prentice Hall, 2000) avoids circular reasoning by defining homologies as "structures... which meet different needs but develop from the same body parts" (pp. 283–284). Guttman's *Biology* (Boston: WCB/McGraw-Hill, 1999) makes a similar claim and throws in genes: "Structures are said to be homologous if they have the same embryonic origins and occupy similar positions in different species.... [and] the information that specifies biological structure is genetic information" (pp. 25, 42). Although these books avoid circular reasoning, they misrepresent the evidence by attributing homologous structures to similar developmental pathways or genes.

Critical thinking in action

The Gee quotations are from Henry Gee, *In Search of Deep Time* (New York: The Free Press, 1999), pp. 9–10.

Chapter 5: Haeckel's Embryos

Darwin quotations (listed in order of their appearance in the text) are from Darwin, *The Origin of Species*, Chapter XIV, pp. 346, 338,

345, 333, 345; *The Descent of Man*, Chapter I, pp. 398, 411. The quotation calling embryology "by far the strongest" evidence is from a September 10, 1860, letter to Asa Gray, in Francis Darwin (editor), *The Life and Letters of Charles Darwin* (New York: D. Appleton & Company, 1896), Vol. II, p. 131; the letter is cited in Ernst Mayr, *The Growth of Biological Thought* (Cambridge, MA: Harvard University Press, 1982), p. 470, and in Stephen Jay Gould, *Ontogeny and Phylogeny* (Cambridge, MA: Harvard University Press, 1977), p. 70.

Figure 5–1: This is the most widely used version of Haeckel's various embryo drawings. It is from Figures 57 & 58 in George J. Romanes, *Darwinism Illustrated* (Chicago: Open Court, 1892), pp. 42–43. It appeared (with black and white reversed) as Tafel VI & VII in Ernst Haeckel, *Anthropogenie, oder Entwicklungsgeschichte des Menschen* (Leipzig: Verlag von Wilhelm Engelmann, 1877), following p. 291; and in English translation as Plates VI & VII in Ernst Haeckel, *The Evolution of Man* (New York: D. Appleton and Company, 1896), following p. 362.

My use of the term "class" here is non-technical and traditional. Cladists object to calling reptiles a class, because they do not consider them a natural group (i.e., one containing a common ancestor and all its descendants). But I'm not a cladist (see the next chapter, on *Archaeopteryx*), and in any case "class" will be more familiar to most readers. The main point of this chapter—that vertebrate embryos are not most similar in their earliest stages—does not depend on the terminology.

Will the real embryologist please stand up?

The quotations of von Baer's laws are from Arthur Henfrey and Thomas H. Huxley (editors), *Scientific Memoirs: Selected from the Transactions of Foreign Academies of Science and from Foreign Journals: Natural History* (London, 1853; reprinted 1966 by Johnson Reprint

Corporation, New York), p. 214. The Lenoir quotation is from Timothy Lenoir, *The Strategy of Life* (Chicago: The University of Chicago Press, 1982), p. 258.

Darwin's misuse of von Baer

Darwin, *The Origin of Species,* pp. 338, 345. On Darwin's misuse of von Baer, see Jane M. Oppenheimer, "An Embryological Enigma in the *Origin of Species*," pp. 221–255 in Jane M. Oppenheimer, *Essays in the History of Embryology and Biology* (Cambridge, MA: The M.I.T. Press, 1967).

The Churchill quotation is from Frederick B. Churchill, "The Rise of Classical Descriptive Embryology," pp. 1–29 in Scott F. Gilbert (editor), *A Conceptual History of Modern Embryology* (Baltimore, MD: The Johns Hopkins University Press, 1991), pp. 19–20.

Haeckel's biogenetic law

Gould, *Ontogeny and Phylogeny,* p. 168; Adam Sedgwick, "The Influence of Darwin on the Study of Animal Embryology," pp. 171–184 in A. C. Seward (editor), *Darwin and Modern Science* (Cambridge: Cambridge University Press, 1909), pp. 174–176; Frank R. Lillie, *The Development of the Chick*, Second Edition (New York: Henry Holt, 1919), p. 6; Gould, *Ontogeny and Phylogeny,* p. 168; Nicholas Rasmussen, "The Decline of Recapitulationism in Early Twentieth-Century Biology: Disciplinary Conflict and Consensus on the Battleground of Theory," *Journal of the History of Biology* 24 (1991), pp. 51–89.

Resurrecting recapitulation

Frank R. Lillie, *The Development of the Chick*, pp. 4–6; Walter Garstang, "The theory of recapitulation: a critical restatement of the biogenetic law," *Journal of the Linnean Society (Zoology),* 35

(1922), pp. 81–101; Gavin de Beer, *Embryos and Ancestors*, Third Edition (Oxford: Clarendon Press, 1958), pp. 10, 164, 172. See also Jane Maienschein, "Cell Lineage, Ancestral Reminiscence, and the Biogenetic Law," *Journal of the History of Biology* 11 (1978), pp. 129–158.

Darwin, *The Origin of Species*, pp. 338, 345. Stephen Jay Gould claims that Darwin never advocated Haeckelian recapitulation, but the plain meaning of Darwin's words belies the claim; see Robert Richards, *The Meaning of Evolution* (Chicago: The University of Chicago Press, 1992), pp. 169–174.

Haeckel's embryo drawings

Jane M. Oppenheimer, "Haeckel's Variations on Darwin," pp. 123–135 in Henry M. Heonigswald and Linda F. Wiener (editors), *Biological Metaphor and Cladistic Classification* (Philadelphia: University of Pennsylvania Press, 1987), p. 134. See also "Accused of Fraud, Haeckel Leaves the Church," *The New York Times* (November 27, 1910), Part 5, p. 11; J. Assmuth and Ernest R. Hull, *Haeckel's Frauds and Forgeries* (Bombay: Examiner Press, 1915); Günter Rager, "Human embryology and the law of biogenesis," *Rivista di Biologia* 79 (1986), pp. 449–465.

In Figure 5–2, the middle line (showing actual embryos) is based largely on data from M. K. Richardson, J. Hanken, M. L. Gooneratne, C. Pieau, A. Raynaud, L. Selwood, and G. M. Wright, "There is no highly conserved embryonic stage in the vertebrates: implications for current theories of evolution and development," *Anatomy & Embryology* 196 (1997), pp. 91–106.

Michael K. Richardson, "Heterochrony and the Phylotypic Period," *Developmental Biology* 172 (1995), pp. 412–421; M. K. Richardson, et al., "There is no highly conserved embryonic stage in the vertebrates: implications for current theories of evolution and development," *Anatomy & Embryology* 196 (1997), pp. 91–106. See also Michael K. Richardson, Steven P. Allen, Glenda M. Wright, Albert Raynaud, and James Hanken, "Somite number and vertebrate evolution," *Development* 125 (1998), pp. 151–160; Elizabeth Pennisi, "Haeckel's Embryos: Fraud Rediscovered," *Science* 277 (1997), p. 1435. Stephen Jay Gould's quote is from his essay, "Abscheulich! (Atrocious!)," *Natural History* (March 2000), pp. 42–49.

The earliest stages in vertebrate embryos are not the most similar

Lewis Wolpert, *The Triumph of the Embryo* (Oxford: Oxford University Press, 1991), p. 12. See also Jonathan Wells, "Haeckel's Embryos and Evolution: Setting the Record Straight," *The American Biology Teacher* 61 (May 1999), pp. 345–349.

In Figure 5–3, the data for earlier stages of zebrafish, frog, chick and human are taken from a variety of standard sources; see Figure 3 in Jonathan Wells, "Haeckel's Embryos and Evolution: Setting the Record Straight," *The American Biology Teacher* 61 (May 1999), pp. 345–349. See also Richard P. Elinson, "Change in developmental patterns: embryos of amphibians with large eggs," pp. 1–21 in R. A. Raff and E. C. Raff (editors), *Development as an Evolutionary Process*, Vol. 8 (New York: Alan R. Liss, 1987).

The data for early turtle development were compiled by Jody F. Sjogren from Louis Agassiz, *Contributions to The Natural History of the United States of America*, First Monograph, Vol. 2 (Boston: Little, Brown and Company, 1857); Oskar Hertwig, *Handbuch der vergleichenden und experimentellen Entwicklungslehre der Wirbeltiere*, Erster Band, Zweiter Teil (Jena: Verlag von Gustav Fischer, 1906); I. Y. Mahmoud, George L. Hess, and John Klicka, "Normal Embryonic Stages of the Western Painted Turtle, *Chrysemys picta bellii*," *Journal of Morphology* 141 (1975), pp. 269–280; Michael A. Ewert,

"Embryology of Turtles," pp. 75–267, and Jeffrey Dean Miller, "Embryology of Marine Turtles," pp. 269–328, in Carl Gans, Frank Billett, and Paul F. A. Maderson (editors), *Biology of the Reptilia*, Vol. 14, Development A (New York: John Wiley & Sons, 1985); S. Renous, F. Rimblot-Baly, J. Fretey, and C. Pieau, "Caractéristiques du développement embryonnaire de la Tortue Luth, *Dermochelys coriacea* (Vandelli, 1761)," *Annales des Sciences Naturelles, Zoologie, Paris*, Series 13, Vol. 10 (1989), pp. 197–229; G. Guyot, C. Pieau, and S. Renous, "Développement embryonnaire d'une tortue terrestre, la tortue d'Hermann, *Testudo hermanni* (Gmelin, 1789)," *Annales des Sciences Naturelles, Zoologie, Paris*, Series 13, Vol. 15 (1994), pp. 115–137; J. J. Pasteels, "Études sur la gastrulation des vertébrés méroblastiques. II. Reptiles," *Archive Biologique, Paris,* 48 (1937), pp. 105–184; J. J. Pasteels. "Études sur la gastrulation des vertébrés méroblastiques. IV. Conclusions generales," *Archive Biologique, Paris,* 48 (1937), pp. 463–488; J. J. Pasteels, "Développement embryonnaire," pp. 893–971 in Pierre P. Grassé, editor, *Traité de Zoologie* (Paris: Masson, 1970), Vol. 14.

In Figure 5–3, the term "placental mammal" for humans is imprecise, because marsupials have a kind of placenta; "eutherian" is the correct term, but it seemed too technical to use here. The main point of the figure is to show that Haeckel's drawings do not fit the actual embryos, whatever they may be called.

The dissimilarity of early embryos is well-known

Adam Sedgwick, "On the Law of Development commonly known as von Baer's Law; and on the Significance of Ancestral Rudiments in Embryonic Development," *Quarterly Journal of Microscopical Science* 36 (1894), pp. 35–52; William W. Ballard, "Problems of gastrulation: real and verbal," *BioScience* 26 (1976), pp. 36–39; Erich Blechschmidt, *The Beginnings of Human Life*, translated by Transemantics

(New York: Springer-Verlag, 1977), pp. 29–30; Richard P. Elinson, "Change in developmental patterns: embryos of amphibians with large eggs," pp. 1–21 in R. A. Raff and E. C. Raff (editors), *Development as an Evolutionary Process*, Vol. 8 (New York: Alan R. Liss, 1987), p. 3; Michael K. Richardson, "Vertebrate evolution: the developmental origins of adult variation," *BioEssays* 21 (1999), pp. 604–613.

On modern terminology for what Haeckel called the "first" stage of development, see William W. Ballard, "Morphogenetic Movements and the Fate Maps of Vertebrates," *American Zoologist* 21 (1981), pp. 391–399, ("pharyngula"); Klaus Sander, "The evolution of patterning mechanisms: gleanings from insect embryogenesis and spermatogenesis," pp. 137–159 in B. C. Goodwin, N. Holder, and C. C. Wylie (editors), *Development and Evolution,* Sixth Symposium of the British Society for Developmental Biology (Cambridge: Cambridge University Press, 1983), p. 140 ("phylotypic stage"); J. M. W. Slack, P. W. H. Holland, and C. F. Graham, "The zootype and the phylotypic stage," *Nature* 361 (1993), pp. 490–492 ("zootype").

Denis Duboule, "Temporal colinearity and the phylotypic progression: a basis for the stability of a vertebrate Bauplan and the evolution of morphologies through heterochrony," *Development* Supplement (1994), pp. 135–142; Michael K. Richardson, "Vertebrate evolution: the developmental origins of adult variation," *BioEssays* 21 (1999), pp. 604–613; Rudolf A. Raff, *The Shape of Life: Genes, Development, and the Evolution of Animal Form* (Chicago: The University of Chicago Press, 1996), p. 197.

Vertebrates are not the only phylum that contradicts von Baer's laws. Recent embryological research on worms, insects, and sea urchins reveals many instances in which organisms in the same group differ more in the early stages of development than in later ones. Von Baer's laws are no more true for invertebrates than for vertebrates. For example, see R. A. Raff, G. Wray, and J. J. Henry, "Implications of radical evolutionary changes in early development for concepts of developmental constraint," pp. 189–207 in L. Warren and H. Koprowski (editors), *New Perspectives in Evolution* (New York: Wiley-Liss, 1991).

A paradox for Darwinian evolution

Gregory Wray, "Punctuated Evolution of Embryos," *Science* 267 (1995), pp. 1115–1116; Raff, *The Shape of Life: Genes, Development, and the Evolution of Animal Form,* p. 211.

Haeckel is dead. Long live Haeckel.

The Balinsky quotation is from B. I. Balinsky, *An Introduction to Embryology*, Fourth Edition (Philadelphia: W. B. Saunders Company, 1975), pp. 7–8 (emphasis in original); Douglas Futuyma, *Evolutionary Biology*, Third Edition (Sunderland, MA: Sinauer Associates, 1998), p. 653; Helena Curtis and N. Sue Barnes, *Invitation to Biology,* Fifth Edition (New York: Worth Publishers, 1994), p. 405; Bruce Alberts, Dennis Bray, Julian Lewis, Martin Raff, Keith Roberts, and James D. Watson, *Molecular Biology of the Cell*, Third Edition (New York: Garland Publishing, 1994) pp. 32–33; Peter H. Raven and George B. Johnson, *Biology,* Fifth Edition (Boston: WCB/McGraw-Hill, 1999), pp. 1181, 416; Cecie Starr and Ralph Taggart, *Biology: The Unity and Diversity of Life,* Eighth Edition (Belmont, CA: Wadsworth Publishing Company, 1998), p. 317; James L. Gould and William T. Keeton (with Carol G. Gould), *Biological Science*, Sixth Edition (New York: W.W. Norton, 1996), p. 347; Burton S. Guttman, *Biology* (Boston: WCB/McGraw-Hill, 1999), p. 718; Sylvia Mader, *Biology*, Sixth Edition (Boston: WCB/McGraw-Hill, 1998), p. 298; Neil A. Campbell, Jane B. Reece, and Lawrence G.

Mitchell, *Biology*, Fifth Edition (Menlo Park, CA: The Benjamin/ Cummings Publishing Company, 1999), p. 424.

Is a human embryo like a fish?

Kenneth Miller, "What Does It Mean To Be One Of Us?" *Life Magazine* (November, 1996), pp. 38–56. Curtis & Barnes, *Invitation to Biology*, p. 405; Gould & Keeton, *Biological Science*, pp. 10, 347; Raven & Johnson, *Biology*, pp. 416, 1181; Futuyma, *Evolutionary Biology*, p. 122.

"Gill slits" are not gill slits

William W. Ballard, "Problems of gastrulation: real and verbal," *Bio-Science* 26 (1976), pp. 36–39; Lewis Wolpert, *The Triumph of the Embryo* (Oxford: Oxford University Press, 1991), p. 185; Günter Rager, "Human embryology and the law of biogenesis," *Rivista di Biologia* 79 (1986), pp. 449–465.

Atrocious!

Douglas Futuyma's remarks were posted February 17, 2000, to the Digital City-Kansas City public evolution board (http://home. digitalcity.com/kansascity); the Gould quotations are from Stephen Jay Gould, "Abscheulich! Atrocious!" *Natural History* (March, 2000), pp. 42–49.

Chapter 6: *Archaeopteryx*: The Missing Link

Darwin, *The Origin of Species*, Chapter X, pp. 235, 234, 255. The discovery of *Archaeopteryx* has been recounted in several recent books on the subject, including Alan Feduccia, *The Origin and Evolution of Birds* (New Haven, CT: Yale University Press, 1996), and Pat Shipman, *Taking Wing* (New York: Simon & Schuster, 1998).

Figure 6–1: This photo of the Berlin *Archaeopteryx* is from Harry G. Seeley, "On some Differences between the London and Berlin Specimens referred to *Archaeopteryx*," *The Geological Magazine*, Series 2, Vol. 8 (1881), pp. 454–455. Photo provided by the Linda Hall Library, Kansas City, Missouri.

The "First Bird"

Lowell Dingus and Timothy Rowe, *The Mistaken Extinction: Dinosaur Evolution and the Origin of Birds* (New York: W. H. Freeman and Company, 1998), p. 116; Alan Feduccia, *The Origin and Evolution of Birds*, p. 29; Pat Shipman, *Taking Wing*, pp. 14–16.

On *Protoavis* see Sankar Chatterjee, "Cranial anatomy and relationships of a new Triassic bird from Texas," *Philosophical Transactions of the Royal Society of London* B 332 (1991), pp. 277–342; Sankar Chatterjee, "*Protoavis* and the early evolution of birds," *Palaeontographica* 254 (1999), pp. 1–100. On the absence of feathers in *Protoavis* fossils, see Roger L. DiSilvestro, "In quest of the origin of birds," *BioScience* 47 (1997), pp. 481–485. Quotations from paleontologists skeptical of *Protoavis* are in Pat Shipman, *Taking Wing*, pp. 112–113. See also Edwin H. Colbert and Michael Morales, *Evolution of the Vertebrates*, Fourth Edition (New York: Wiley-Liss, 1991), p. 183; Alan Feduccia, *The Origin and Evolution of Birds*, p. 38.

For the accusation that *Archaeopteryx* was a forgery, see Fred Hoyle and Chandra Wickramasinghe, *Archaeopteryx, the Primordial Bird: A Case of Fossil Forgery* (London: Christopher Davies, 1986). For the refutation of this claim, see Alan J. Charig et al., "*Archaeopteryx* Is Not a Forgery," *Science* 232 (1986), pp. 622–626; David Dickson, "Feathers Still Fly in Row over Fossil Bird," *Science* 238 (1987), pp. 475–476; Giles Courtice, "Museum officials confident *Archaeopteryx* is genuine... but opponents renew demands for proof," *Nature* 328 (1987), p. 657. See also Peter Wellnhofer, "*Archaeopteryx*," *Scientific American* 262 (May, 1990), pp. 70–77; Feduccia, *The Origin and Evolution of Birds*, pp. 38–39; Shipman, *Taking Wing*, pp. 141–148.

The missing link

On the original dinosaur theory of bird evolution see Thomas H. Huxley, "On the Animals which are most nearly intermediate between Birds and Reptiles," *The Annals and Magazine of Natural History*, Vol. II, Fourth Series (1868), pp. 66–75; Darwin, *Origin of Species*, p. 266. See also John H. Ostrom, "*Archaeopteryx* and the origin of birds," *Biological Journal of the Linnean Society* 8 (1976), pp. 91–182; Adrian Desmond, *Archetypes and Ancestors* (Chicago: The University of Chicago Press, 1982), pp. 124–131. On the misidentification of *Archaeopteryx* as *Compsognathus*, see Dingus & Rowe, *The Mistaken Extinction*, pp. 120, 185; Shipman, *Taking Wing*, pp. 44–45, 115.

Coelophysis, a two-legged dinosaur that preceded *Archaeopteryx,* is not considered ancestral to *Archaeopteryx* because, like *Compsognathus,* its features are not those one would expect in an ancestor; see Robert L. Carroll, *Vertebrate Paleontology and Evolution* (New York: W. H. Freeman, 1988), pp. 290–292, 303; and Dingus and Rowe, *The Mistaken Extinction*, pp. 181–183.

The Mayr quotation is from Ernst Mayr, *The Growth of Biological Thought* (Cambridge, MA: Harvard University Press, 1982), p. 430.

Quotations about *Archaeopteryx* not being the ancestor of modern birds are from Larry D. Martin, "The Relationship of *Archaeopteryx* to other Birds," pp. 177–183 in M. K. Hecht, J. H. Ostrom, G. Viohl, and P. Wellnhofer (editors), *The Beginnings of Birds* (Eichstätt: Freunde des Jura-Museums, 1985), p. 182;

John Schwartz, "New Evolution Research Ruffles Some Feathers," *The Washington Post* (November 15, 1996), p. A3 (quoting Mark Norell).

The origin of flight

For discussions of the "trees down" and "ground up" theories of the origin of flight, see Walter J. Bock, "The Arboreal Origin of Avian Flight," pp. 57–72, and John H. Ostrom, "The Cursorial Origin of Avian Flight," pp. 73–81, in Kevin Padian (editor), *The Origin of Birds and the Evolution of Flight* (San Francisco: California Academy of Sciences, 1986), Memoir Number 8; Feduccia, *The Origin and Evolution of Birds*, pp. 93–137; Shipman, *Taking Wing*, pp. 174–218.

Cladistics

Kevin de Queiroz, "Systematics and the Darwinian Revolution," *Philosophy of Science* 55 (1988), pp. 238–259. See also Kevin de Queiroz and Jacques Gauthier, "Toward a phylogenetic system of biological nomenclature," *Trends in Ecology and Evolution* 9 (1994), pp. 27–31; Henry Gee, *In Search of Deep Time* (New York: The Free Press, 1999). On the application of cladistics to bird phylogeny, see Jacques Gauthier, "Saurischian Monophyly and the Origin of Birds," pp. 1–55, in Kevin Padian (editor), *The Origin of Birds and the Evolution of Flight*. The Shipman quotation is from *Taking Wing*, p. 33.

Re-arranging the evidence

The data for Figure 6–2 (Cladistic theory and the fossil record) are from Kevin Padian and Luis M. Chiappe, "The origin and early evolution of birds," *Biological Reviews* 73 (1998), pp. 1–42 (Figure 14). The figure had been previously published in Luis M. Chappe,

"The first 85 million years of avian evolution," *Nature* 378 (1995), pp. 349–355.

Quotations by Chiappe and Ruben are from Roger L. DiSilvestro, "In quest of the origin of birds," *BioScience* 47 (1997), pp. 481–485.

Dethroning Archaeopteryx

Birds are dinosaurs: Dingus and Rowe, *The Mistaken Extinction*, pp. 205–206. See also Kevin Padian and Luis M. Chiappe, "The Origin of Birds and Their Flight," *Scientific American* (February, 1998), pp. 38–47.

Gee quotations are from Henry Gee, *In Search of Deep Time* (New York: The Free Press, 1999), pp. 195–197.

The *"Piltdown bird"*

Christopher P. Sloan, "Feathers for T. Rex?" *National Geographic* 196 (November, 1999), pp. 98–107. National Geographic's website retraction is at http://www.ngnews.com/ news/2000/01/ 01212000/feathereddino_9321.txt. Rex Dalton, "Feathers fly over Chinese fossil bird's legality and authenticity," *Nature* 403 (2000), pp. 689–690; "Fossil smuggling unopposed," *Nature* 403 (2000), p. 687; William L. Allen, "Fooled, but not foolish," letter to *Nature* 404 (2000), p. 541; Xu Xing, "Feathers for T. rex?" letter to *National Geographic* (March, 2000), pp. Forum section. See also Constance Holden, "Florida Meeting Shows Perils, Promise of Dealing for Dinos," *Science* 288 (2000), pp. 238–239; Jeff Hecht, "Piltdown bird," *New Scientist* 165 (January 29, 2000), p. 12; Rex Dalton, "Fake bird fossil highlights the problem of illegal trading," *Nature* 404 (2000), p. 696.

The open letter from Storrs Olson to Peter Raven was dated November 1, 1999, and sent in eletronic form (as an email message) and hard copy. The authenticity of the letter and its contents were confirmed to me in a personal communication from Storrs Olson on April 24, 2000.

Feathers for Bambiraptor

The Florida Symposium on Dinosaur Bird Evolution, April 7 and 8, 2000, Ft. Lauderdale, Florida. Sponsored by the Florida Institute of Paleontology and The Graves Museum of Archaeology and Natural History. The original scientific description is David A. Burnham, Kraig L. Derstler, Philip J. Currie, Robert T. Bakker, Zhonghe Zhou, and John H. Ostrom, "Remarkable New Birdlike Dinosaur (Theropoda: Maniraptora) from the Upper Cretaceous of Montana," *The University of Kansas Paleontological Contributions*, New Series, Number 13 (March 15, 2000). Paleontologists' quotes are from "Another Birdlike Dino Unveiled," *Science* 287 (March 24, 2000), p. 2145; and David Burnham, in a videotape played in the exhibit room at the Florida conference, April 7–8, 2000. See also Constance Holden, "Florida Meeting Shows Perils, Promise of Dealing for Dinos," *Science* 288 (2000), pp. 238–239.

Figures 6–3 and 6–4 are based on the reconstructed *Bambiraptor* specimen displayed at the Florida Symposium on Dinosaur Bird Evolution, April 7–8, 2000. See also David A. Burnham, Kraig L. Derstler, Philip J. Currie, Robert T. Bakker, Zhonghe Zhou, and John H. Ostrom, "Remarkable New Birdlike Dinosaur (Theropoda: Maniraptora) from the Upper Cretaceous of Montana," *The University of Kansas Paleontological Contributions*, New Series, Number 13 (March 15, 2000).

On feathered *Velociraptors* at the American Museum of Natural History in New York, see Sharon Begley and Thomas Hayden, "When Dinsoaurs Roamed the Earth," *Newsweek* (May 15, 2000), pp. 66–68.

The quotations from Feduccia and Martin are from Pat Shipman, "Birds do it... did dinosaurs?" *New Scientist* (February 1, 1997), pp. 27–31.

Turkey DNA from Triceratops?

Damien Marsic, Parker Carroll, Laura Heffelfinger, Tyler Lyson, Joseph D. Ng, and William R. Garstka, "DNA Sequence of the Mitochondrial 12S rRNA Gene from *Triceratops* Fossils: Molecular Evidence Supports the Evolutionary Relationship between Dinosaurs and Birds," *Publications in Paleontology*, No. 2, Graves Museum of Archaeology and Natural History, Dania Beach, FL (April 7–8, 2000), p. 19; Constance Holden, "Dinos and Turkeys: Connected by DNA?" *Science* 288 (2000), p. 238.

On the inability to recover useful sequence information from DNA older than a million years, see Tomas Lindahl, "Instability and decay of the primary structure of DNA," *Nature* 362 (1993), pp. 709–715.

The "cracked kettle" approach to doing science

The title of Kevin Padian's talk was "Methods and Standards of Evidence: Why the Bird-Dinosaur Controversy is Dead." The abstract is in *Publications in Paleontology*, No. 2, Graves Museum of Archaeology and Natural History, Dania Beach, FL (April 7–8, 2000), p. 21.

Whatever happened to Archaeopteryx?

Textbook quotations are from Sylvia Mader, *Biology*, Sixth Edition (Boston, MA: WCB/McGraw-Hill, 1998), p. 296; William D. Schraer and Herbert J. Stoltze, *Biology: The Study of Life*, Seventh Edition (Upper Saddle River, NJ: Prentice Hall, 1999), p. 761.

Chapter 7: Peppered Moths

Darwin, *The Origin of Species*, Introduction, p. 14; the "imaginary illustrations" quotation is from Chapter IV, p. 70; Hermon C. Bumpus, "The Elimination of the Unfit as Illustrated by the Introduced Sparrow, *Passer domesticus*," pp. 209–226 in *Biological Lectures from the Marine Biological Laboratory, 1898* (Boston: Ginn & Company, 1899). See also John Endler, *Natural Selection in the Wild* (Princeton, NJ: Princeton University Press, 1986); Jonathan Weiner, *The Beak of the Finch* (New York: Vintage Books, 1994), pp. 226–227; Ernst Mayr, *The Growth of Biological Thought* (Cambridge, MA: Harvard University Press, 1982), p. 586; H. B. D. Kettlewell, "Darwin's Missing Evidence," *Scientific American* 200 (March 1959), pp. 48–53.

Industrial melanism

Most acounts of industrial melanism claim that the first melanic moth was captured in 1848, but several writers refer to a collection made before 1811; see E. B. Ford, *Ecological Genetics*, Fourth Edition (London: Chapman and Hall, 1975), p. 329.

   J. W. Tutt, *British Moths* (London: George Routledge, 1896); J. W. H. Harrison, "Genetical studies in the moths of the geometrid genus *Oporabia (Oporinia)* with a special consideration of melanism in the Lepidoptera," *Journal of Genetics* 9 (1920), pp. 195–280; J. W. Heslop Harrison, "The Experimental Induction of Melanism, and other Effects, in the Geometrid Moth *Selenia bilunaria* esp.," *Proceedings of the Royal Society of London* B 117 (1935), pp. 78–92; E. B. Ford, "Problems of heredity in the Lepidoptera," *Biological Reviews* 12 (1937), pp. 461–503; E. B. Ford, *Ecological Genetics*, pp. 319–321. See also Michael E. N. Majerus, *Melanism: Evolution in Action* (Oxford: Oxford University Press, 1998).

Kettlewell's experiments

H. B. D. Kettlewell, "Selection experiments on industrial melanism in the Lepidoptera," *Heredity* 9 (1955), pp. 323–342; H. B. D. Kettlewell, "Further selection experiments on industrial melanism in the Lepidoptera," *Heredity* 10 (1956), pp. 287–301. See also Bernard Kettlewell, *The Evolution of Melanism* (Oxford: Clarendon Press, 1973).

Darwin's missing evidence

Quotations are from H. B. D. Kettlewell, "Selection experiments on industrial melanism in the Lepidoptera," *Heredity* 9 (1955), pp. 323–342; H. B. D. Kettlewell, "Darwin's Missing Evidence," *Scientific American* 200 (March 1959), pp. 48–53; P. M. Sheppard, *Natural Selection and Heredity*, Fourth Edition (London: Hutchinson University Library, 1975), p. 70; Sewall Wright, *Evolution and the Genetics of Populations*, Vol. 4: Variability Within and Among Natural Populations (Chicago: The University of Chicago Press, 1978), p. 186; J. S. Jones, "More to melanism than meets the eye," *Nature* 300 (1982), p. 109.

On the decline of melanism see C. A. Clarke and P. M. Sheppard, "A local survey of the distribution of industrial melanic forms in the moth *Biston betularia* and estimates of the selective values of these in an industrial environment," *Proceedings of the Royal Society of London* B 165 (1966), pp. 424–439; Bernard Kettlewell, *The Evolution of Melanism;* J. A. Bishop and Laurence M. Cook, "Moths, Melanism and Clean Air," *Scientific American* 232 (1975), pp. 90–99. See also D. R. Lees, "Industrial melanism: genetic adaptation of animals to air pollution," pp. 129–176 in J. A. Bishop and L. M. Cook (editors), *Genetic Consequences of Man-made Change* (London: Academic Press, 1981).

Problems with the evidence

J. A. Bishop, "An experimental study of the cline of industrial melanism in *Biston betularia* (L.) (Lepidoptera) between urban Liverpool and rural North Wales," *Journal of Animal Ecology* 41 (1972), pp. 209–243; D. R. Lees and E. R. Creed, "Industrial melanism in *Biston betularia:* the role of selective predation," *Journal of Animal Ecology* 44 (1975), pp. 67–83; R. C. Steward, "Industrial and non-industrial melanism in the peppered moth, *Biston betularia* (L.)," *Ecological Entomology* 2 (1977), pp. 231–243; R. J. Berry, "Industrial melanism and peppered moths (*Biston betularia* (L.))," *Biological Journal of the Linnean Society* 39 (1990), pp. 301–322. See also J. A. Bishop and L. M. Cook, "Industrial melanism and the urban environment," *Advances in Ecological Research* 11 (1980), pp. 373–404; G. S. Mani, "Theoretical models of melanism in *Biston betularia*—a review," *Biological Journal of the Linnean Society* 39 (1990), pp. 355–371.

The exaggerated role of lichens

Bernard Kettlewell, *The Evolution of Melanism*; D. R. Lees, E. R. Creed, and L. G. Duckett, "Atmospheric pollution and industrial melanism," *Heredity* 30 (1973), pp. 227–232; C. A. Clarke, G. S. Mani, and G. Wynne, "Evolution in reverse: clean air and the peppered moth," *Biological Journal of the Linnean Society* 26 (1985), pp. 189–199; Bruce S. Grant and Rory J. Howlett, "Background selection by the peppered moth (*Biston betularia* Linn.): individual differences," *Biological Journal of the Linnean Society* 33 (1988), pp. 217–232; B. S. Grant, D. F. Owen, and C. A. Clarke, "Parallel Rise and Fall of Melanic Peppered Moths in America and Britain," *Journal of Heredity* 87 (1996), pp. 351–357; B. S. Grant, A. D. Cook, C. A. Clarke, and D. F. Owen, "Geographic and Temporal Variation in the Incidence of Melanism in Peppered Moth Populations in America and Britain," *Journal of Heredity* 89 (1998), pp. 465–471. See also D. F. Owen, "The Evolution of Melanism in Six Species of North American Geometrid Moths," *Annals of the Entomological Society of America* 55 (1962), pp. 695–703; Bruce S. Grant, Denis F. Owen, and Cyril A. Clarke, "Decline of melanic moths," *Nature* 373 (1995), p. 565.

Peppered moths don't rest on tree trunks

The one attempt to release moths before dawn is described in Bernard Kettlewell, *The Evolution of Melanism*, p. 129; Kettlewell's quotation about moths choosing positions higher in the trees is from H. B. D. Kettlewell, "Selection experiments on industrial melanism in the Lepidoptera," *Heredity* 9 (1955), pp. 323–342.

Research using dead specimens glued or pinned to tree trunks included C. A. Clarke and P. M. Sheppard, "A local survey of the distribution of industrial melanic forms in the moth *Biston betularia* and estimates of the selective values of these in an industrial environment," *Proceedings of the Royal Society of London* B 165 (1966), pp. 424–439; J. A. Bishop, "An experimental study of the cline of industrial melanism in *Biston betularia* (L.) (Lepidoptera) between urban Liverpool and rural North Wales," *Journal of Animal Ecology* 41 (1972), pp. 209–243; D. R. Lees and E. R. Creed, "Industrial melanism in *Biston betularia:* the rôle of selective predation," *Journal of Animal Ecology* 44 (1975), pp. 67–83; R. C. Steward, "Melanism and selective predation in three species of moths," *Journal of Animal Ecology* 46 (1977), pp. 483–496; N. D. Murray, J. A. Bishop, and M. R. MacNair, "Melanism and predation by birds in the moths *Biston betularia* and *Phigalia pilosauria,*" *Proceedings of the Royal Society of London* B 210 (1980), pp. 277–283.

Misgivings about the use of dead moths were expressed by Kettlewell in his book, *The Evolution of Melanism*, p. 150; and by J. A. Bishop and Laurence M. Cook, "Moths, Melanism and Clean Air," *Scientific American* 232 (1975), pp. 90–99.

For actual evidence regarding the moth's natural resting places, see K. Mikkola, "On the selective forces acting in the industrial melanism of *Biston* and *Oligia* moths (Lepidoptera: Geometridae and Noctuidae)," *Biological Journal of the Linnean Society* 21 (1984), pp. 409–421; C. A. Clarke, G. S. Mani, and G. Wynne, "Evolution in reverse: clean air and the peppered moth," *Biological Journal of the Linnean Society* 26 (1985), pp. 189–199; Rory J. Howlett and Michael E. N. Majerus, "The understanding of industrial melanism in the peppered moth (*Biston betularia*) (Lepidoptera: Geometridae)," *Biological Journal of the Linnean Society* 30 (1987), pp. 31–44; Tony G. Liebert and Paul M. Brakefield, "Behavioural studies on the peppered moth *Biston betularia* and a discussion of the role of pollution and lichens in industrial melanism," *Biological Journal of the Linnean Society* 31 (1987), pp. 129–150; M. E. N. Majerus, *Melanism: Evolution in Action*, p. 116. For a short review, see Jeremy Cherfas, "Exploding the myth of the melanic moth," *New Scientist* (December 25, 1986–January 1, 1987), p. 25.

Staged photographs

A 1975 photo using torpid live moths is in J. A. Bishop and Laurence M. Cook, "Moths, Melanism and Clean Air," *Scientific American* 232 (1975), pp. 90–99. (The procedure for making the photo was confirmed to me in a personal communication from L. M. Cook, 1998, University of Manchester, Manchester, U.K.) The Sargent statement is from Larry Witham, "Darwinism icons disputed: Biologists discount moth study," *The Washington Times* (National Weekly Edition) (January 25–31, 1999), p. 28.

Doubts about the classical story

Giuseppe Sermonti and Paola Catastini, "On industrial melanism: Kettlewell's missing evidence," *Rivista di Biologia* 77 (1984), pp. 35–52; Atuhiro Sibatani, "Industrial Melanism Revisited," *Rivista di Biologia* 92 (1999), pp. 349–356. See also David M. Lambert, Craig D. Millar, and Tony G. Hughes, "On the classic case of natural selection," *Rivista di Biologia* 79 (1986), pp. 11–49; Craig Millar and David Lambert, "Industrial melanism—a classic example of another kind?" a review of Michael Majerus's *Melanism: Evolution in Action*, *BioScience* 49 (1999), pp. 1021–1023.

Theodore D. Sargent, Craig D. Millar, and David M. Lambert, "The 'Classical' Explanation of Industrial Melanism: Assessing the Evidence," *Evolutionary Biology* 30 (1998), pp. 299–322. See also Jonathan Wells, "Second Thoughts about Peppered Moths," *The Scientist* (May 24, 1999), p. 13.

On thermal melanism in ladybird beetles, see E. R. Creed, "Geographic variation in the two-spot ladybird in England and Wales," *Heredity* 21 (1966), pp. 57–72; Paul M. Brakefield, "Polymorphic Müllerian mimicry and interactions with thermal melanism in ladybirds and a soldier beetle: a hypothesis," *Biological Journal of the Linnean Society* 26 (1985), pp. 243–267. See also E. B. Ford, *Ecological Genetics*.

For recent defenses of the classical story that acknowledge its complexities, see Michael E. N. Majerus, *Melanism: Evolution in Action* (Oxford: Oxford University Press, 1998); Laurence Cook, a review of Michael Majerus's *Melanism: Evolution in Action*, *Genetical Research, Cambridge* 72 (1998), pp. 73–75; M. E. N. Majerus, C. F. A. Brunton, and J. Stalker, "A bird's eye view of the peppered moth," *Journal of Evolutionary Biology* 13 (2000), pp. 155–159; L. M. Cook, "Changing views on melanic moths," *Biological Journal of the Linnean Society* 69 (2000), pp. 431–441.

Science or alchemy?

Jerry Coyne, "Not black and white," a review of Michael Majerus's *Melanism: Evolution in Action, Nature* 396 (1998), pp. 35–36; Bruce S. Grant, "Fine Tuning the Peppered Moth Paradigm," *Evolution* 53 (1999), pp. 980–984; John A. Endler, *Natural Selection in the Wild* (Princeton, NJ: Princeton University Press, 1986), p. 164.

The peppered myth

Textbook quotations are from Kenneth R. Miller and Joseph Levine, *Biology*, Fifth Edition (Upper Saddle River, NJ: Prentice Hall, 2000), pp. 297, 298; Burton S. Guttman, *Biology* (Boston, MA: WCB/McGraw-Hill, 1999), pp. 35–36; George B. Johnson, *Biology: Visualizing Life,* Annotated Teacher's Edition (Orlando, FL: Holt, Rinehart and Winston, 1998), p. 182; Cecie Starr and Ralph Taggart, *Biology: The Unity and Diversity of Life*, Eighth Edition (Belmont, CA: Wadsworth Publishing, 1998), p. 286.

One laudable exception to the widespread use of staged pictures is Mark Ridley's textbook, *Evolution*, Second Edition (Cambridge, MA: Blackwell Science, 1996), which carries photos of peppered moths resting under horizontal branches on p. 104.

Other textbooks that feature the peppered myth include: Kenneth R. Miller and Joseph Levine, *Biology: The Living Science* (Upper Saddle River, NJ: Prentice Hall, 1998), p. 234; Eric Strauss and Marilyn Lisowski, *Biology:The Web of Life*, Second Edition (Glenview, IL: Scott Foresman/Addison Wesley, 2000), p. 250; Sylvia Mader, *Biology*, Sixth Edition (Boston, MA:WCB/McGraw-Hill, 1998), p. 310; Teresa Audesirk and Gerald Audesirk, *Biology: Life on Earth*, Fifth Edition (Upper Saddle River, NJ: Prentice Hall, 1999), p. 268.

The Ritter quotation is from Carla Yu, "Moth-eaten Darwinism: A disproven textbook case of natural selection refuses to die," *Alberta Report Newsmagazine* Vol. 26, No. 15 (April 5, 1999), pp. 38–39. The textbook in question is Bob Ritter, Richard F. Coombs, R. Bruce Drysdale, Grant A. Gardner, and Dave T. Lunn, *Biology* (Scarborough, ONT: Nelson Canada, 1993), which deals with peppered moths on pp. 109–110.

The Coyne quotations are from Jerry Coyne, "Not black and white," a review of Michael Majerus's *Melanism: Evolution in Action,* *Nature* 396 (1998), pp. 35–36.

Chapter 8: Darwin's Finches

The legend of Darwin's finches

Sulloway's quotations are from Frank J. Sulloway, "Darwin and His Finches: The Evolution of a Legend," *Journal of the History of Biology* 15 (1982), pp. 1–53; Sulloway, "Darwin and the Galápagos," *Biological Journal of the Linnean Society* 21 (1984), pp. 29–59. See also Sulloway, "Darwin's Conversion: The *Beagle* Voyage and Its Aftermath," *Journal of the History of Biology* 15 (1982), pp. 325–396; and Sulloway, "The legend of Darwin's finches," letter to *Nature* 303 (1983), p. 372.

The first edition of Darwin's journal mentions that "in the thirteen species of ground-finches, a nearly perfect gradation may be traced, from a beak extraordinarily thick, to one so fine, that it may be compared to that of a warbler. I very much suspect, that certain members of the series are confined to different islands; therefore, if the collection had been made on any *one* island, it would not have presented so perfect a gradation." Charles Darwin, *Journal of Researches into the Geology and Natural History of the various countries visited by H. M. S. Beagle,* (1839), Facsimile Reprint of the First Edition (New York: Hafner Publishing, 1952), p. 475. The fact that Darwin referred to "thirteen" species, the number currently recognized, is mere coincidence; his thirteen "species" are not the modern thirteen. The expanded quotation from the second edition of the journal is from Charles Darwin, *Journal of Researches into the Natural History and Geology of the Countries Visited during the Voyage of H. M. S. Beagle Round the World, under the Command of Capt. Fitz-Roy, R. N.,* Second Edition (London: John Murray, 1845), p. 380.

The finches were first called "Darwin's" in Percy Lowe, "The Finches of the Galápagos in relation to Darwin's Conception of Species," *Ibis* 6 (1936), pp. 310–321. The name was popularized by David Lack, *Darwin's Finches* (Cambridge: Cambridge University Press, 1947).

Darwin's finches as an icon of evolution

Sulloway quotations are from Frank J. Sulloway, "Darwin and His Finches: The Evolution of a Legend," *Journal of the History of Biology* 15 (1982), pp. 1–53.

Textbook selections are from James L. Gould and William T. Keeton, *Biological Science*, Sixth Edition (New York: W. W. Norton, 1996), p. 500; Peter H. Raven and George B. Johnson, *Biology*, Fifth Edition (Boston: WCB/McGraw-Hill, 1999), p. 410; George B. Johnson, *Biology: Visualizing Life,* Annotated Teacher's Edition (Orlando, FL: Holt, Rinehart & Winston, 1998), p. 174.

Evidence for evolution?

On the genetics of finch beaks: There have been several studies on the heritability of beaks, meaning the likelihood that offspring will closely resemble their parents in this trait. Heritability of beak depth in *Geospiza fortis*, the species most intensely studied by the Grants, is about 80 percent. Although this may point to a strong genetic (i.e., DNA-encoded) component, it does not identify which genes might be involved. See Peter T. Boag, "The Heritability of External Morphology in Darwin's Ground Finches (*Geospiza*) on Isla

Daphne Major, Galápagos," *Evolution* 37 (1983), pp. 877–894; Peter R. Grant, *Ecology and Evolution of Darwin's Finches* (Princeton: Princeton University Press, 1986), pp. 180–182; Peter T. Boag and Arie J. van Noordwijk, "Quantitative Genetics," pp. 45–78 in F. Cooke and P. A. Buckley (editors), *Avian Genetics: A Population and Ecological Approach* (London: Academic Press, 1987).

A 1984 genetic study of Darwin's finches found little or no genetic difference among several species; see James L. Patton, "Genetical processes in the Galápagos," *Biological Journal of the Linnean Society* 21 (1984), pp. 91–111. A 1984 study which did not identify any genes involved in determining beak morphology was T. D. Price, P. R. Grant, and P. T. Boag, "Genetic Changes in the Morphological Differentiation of Darwin's Ground Finches," pp. 49–66 in K. Wöhrmann and V. Loeschcke (editors), *Population Biology and Evolution* (Berlin: Springer-Verlag, 1984). See also Peter R. Grant, *Ecology and Evolution of Darwin's Finches*, pp. 177, 198–199, 281–283, 395, 399, 405–406.

On the lack of observable chromosome differences among the finches, see Nancy Jo, "Karyotypic Analysis of Darwin's Finches," pp. 201–217 in Robert I. Bowman, Margaret Berson, and Alan E. Leviton (editors), *Patterns of Evolution in Galápagos Organisms* (San Francisco, CA: Pacific Division, AAAS, 1983).

There have been a number of molecular phylogenetic studies of Darwin's finches, but molecular phylogeny must, by its very nature, rely on genes that are not subject to natural selection—otherwise mutations would not accumulate merely as a function of time and DNA sequence differences would not reflect divergence times. For some recent studies, see Kenneth Petren, B. Rosemary Grant, and Peter R. Grant, "A phylogeny of Darwin's finches based on microsatellite DNA length variation," *Proceedings of the Royal Society of London* B 266 (1999), pp. 321–329; Akie Sato, Colm O'hUigin,

Felipe Figueroa, Peter R. Grant, B. Rosemary Grant, Herbert Tichy, and Jan Klein, "Phylogeny of Darwin's finches as revealed by mtDNA sequences," *Proceedings of the National Academy of Sciences USA* 96 (1999), pp. 5101–5106.

Peter and Rosemary Grant concluded in 1997: "The knowledge base from which to generalize about the genetics of bird speciation is precariously thin." Peter R. Grant and B. Rosemary Grant, "Genetics and the origin of bird species," *Proceedings of the National Academy of Sciences USA* 94 (1997), pp. 7768–7775.

The beak of the finch

For the details of this famous story see Jonathan Weiner, *The Beak of the Finch* (New York: Vintage Books, 1994); the quotations are from pp. 9, 112. The Grant quotation is from Peter R. Grant, "Natural Selection and Darwin's Finches," *Scientific American* 265 (October 1991), pp. 82–87. See also Peter T. Boag and Peter R. Grant, "Intense Natural Selection in a Population of Darwin's Finches (Geospizinae) in the Galápagos," *Science* 214 (1981), pp. 82–85; Peter R. Grant, *Ecology and Evolution of Darwin's Finches* (Princeton: Princeton University Press, 1986).

When the rains returned

Quotations about the reversal of selection are from H. Lisle Gibbs and Peter R. Grant, "Oscillating selection on Darwin's finches," *Nature* 327 (1987), pp. 511–513; Weiner, *The Beak of the Finch*, pp. 104–105, 176; Peter R. Grant, "Natural Selection and Darwin's Finches," *Scientific American* 265 (October 1991), pp. 82–87. See also Peter R. Grant, *Ecology and Evolution of Darwin's Finches*, pp. 184, 375, 395; Peter R. Grant and B. Rosemary Grant, "Predicting Microevolutionary Responses to Directional Selection on Heritable Variation," *Evolution* 49 (1995), pp. 241–251.

A reversal of drought-induced selection after the rains returned was also observed in the large cactus finch on Isla Genovesa; see B. Rosemary Grant and Peter R. Grant, *Evolutionary Dynamics of a Natural Population* (Chicago: The University of Chicago Press, 1989).

According to paleobiologist Robert Carroll, oscillating natural selection is the rule rather than the exception. See Robert L. Carroll, "Towards a new evolutionary synthesis," *Trends in Ecology and Evolution* 15 (2000), pp. 27–32: "Over the duration of most species, the intensity and direction of selection change repeatedly, either in an oscillating manner or in what appears to be a random walk.... for much of the duration of the majority of species there is relatively little net change, even over hundreds of thousands of years."

Diverging or merging?

The Grant quotations are from B. Rosemary Grant and Peter R. Grant, "Evolution of Darwin's finches caused by a rare climatic event," *Proceedings of the Royal Society of London* B 251 (1993), pp. 111–117; Peter R. Grant and B. Rosemary Grant, "Hybridization of Bird Species," *Science* 256 (1992), pp. 193–197. Weiner's quotations are from his *The Beak of the Finch*, pp. 197, 176.

See also B. Rosemary Grant and Peter R. Grant, "High Survival of Darwin's Finch Hybrids: Effects of Beak Morphology and Diets," *Ecology* 77 (1996), pp. 500–509; B. Rosemary Grant and Peter R. Grant, "Hybridization and Speciation in Darwin's Finches," pp. 404–422, in Daniel J. Howard and Stewart H. Berlocher (editors), *Endless Forms: Species and Speciation* (New York: Oxford University Press, 1998); Peter R. Grant and B. Rosemary Grant, "Speciation and hybridization of birds on islands," pp. 142–162, in Peter R. Grant (editor), *Evolution on Islands* (Oxford: Oxford University Press, 1998).

Fourteen species, or six?

The Grant quotations are from Peter R. Grant and B. Rosemary Grant, "Hybridization of Bird Species," *Science* 256 (1992), pp. 193–197; Peter R. Grant, "Hybridization of Darwin's finches on Isla Daphne Major, Galápagos," *Philosophical Transactions of the Royal Society of London* B 340 (1993), pp. 127–139. See also P. Grant, *Ecology and Evolution of Darwin's Finches*, p. 206; B. Rosemary Grant and Peter R. Grant, "Hybridization and Speciation in Darwin's Finches," pp. 404–422, in Daniel J. Howard and Stewart H. Berlocher (editors), *Endless Forms: Species and Speciation* (New York: Oxford University Press, 1998).

Exaggerating the evidence

The Grant quotations are from Peter R. Grant and B. Rosemary Grant, "Speciation and hybridization in island birds," *Philosoph-ical Transactions of the Royal Society of London* B 351 (1996), pp. 765–772; Peter R. Grant and B. Rosemary Grant, "Speciation and hybridization of birds on islands," pp. 142–162 in Peter R. Grant (editor), *Evolution on Islands* (Oxford: Oxford University Press, 1998), p. 155. The Ridley quotation is from Mark Ridley, *Evolution*, Second Edition (Cambridge, MA: Blackwell Science, 1996), pp. 570–571.

National Academy of Sciences, *Science and Creationism: A View from the National Academy of Sciences*, Second Edition (Washington, DC: National Academy of Sciences Press, 1999), Chapter on "Evidence Supporting Biological Evolution," p. 2; except for the "compelling example of speciation" hyperbole, the same story was presented in the National Academy's booklet, *Teaching About Evolution and the Nature of Science* (Washington, DC: National Academy Press, 1998), Chapter 2, p. 10. The Johnson quotation is from Phillip E. Johnson, "The Church of Darwin," *The Wall Street Journal* (August 16, 1999), pp. A14.

313 · Research Notes

Chapter 9: Four-Winged Fruit Flies

Figure 9–1 based on E. B. Lewis, "Control of Body Segment Differentation in *Drosophila* by the Bithorax Gene Complex," pp. 269–288 in Max M. Burger and Rudolf Weber (editors), *Embryonic Development, Part A: Genetic Aspects* (New York, Alan R. Liss, 1982), Fig. 3, p. 274.

The origin of variations from Darwin to DNA

Darwin, *The Origin of Species,* Chapter I, p. 37. See also Ernst Mayr, *The Growth of Biological Thought: Diversity, Evolution and Inheritance* (Cambridge, MA: Harvard University Press, 1982); Peter J. Bowler, *Evolution: The History of an Idea*, Second Edition (Berkeley, CA: University of California Press, 1989). The Dobzhansky quotation is from Theodosius Dobzhansky, *Genetics and the Origin of Species* (New York: Columbia University Press, 1937), p. 13. The Monod quotation is from Horace Freeland Judson, *The Eighth Day of Creation: The Makers of the Revolution in Biology* (New York: Simon and Schuster, 1979), p. 217.

Beneficial biochemical mutations

For an introduction to the enormous literature on antibiotic resistance, see Harold C. Neu, "The Crisis in Antibiotic Resistance," *Science* 257 (1992), pp. 1064–1073; Julian Davies, "Inactivation of Antibiotics and the Dissemination of Resistance Genes," *Science* 264 (1994), pp. 375–382; Brian G. Spratt, "Resistance to Antibiotics Mediated by Target Alterations," *Science* 264 (1994), pp. 388–393; Martin C. J. Maiden, "Horizontal Genetic Exchange, Evolution, and Spread of Antibiotic Resistance in Bacteria," *Clinical Infectious Diseases* 27 Supplement 1 (1998): S12–S20.

On enzymatic inactivation as the most common cause of insecticide resistance, see Michel Raymond, Amanda Callaghan, Phillipe

Fort, and Nicole Pasteur, "Worldwide migration of amplified insecticide resistance genes in mosquitoes," *Nature* 350 (1991), pp. 151–153. For some general background on the role of mutations in insecticide and pesticide resistance, see Richard T. Roush and John A. McKenzie, "Ecological Genetics of Insecticide and Acaricide Resistance," *Annual Review of Entomology* 32 (1987), pp. 361–380.

On sickle-cell anemia, see Anthony C. Allison, "Sickle Cells and Evolution," *Scientific American* 195 (1956), pp. 87–94; F. Vogel and A. G. Motulsky, *Human Genetics*, Third Edition (Berlin: Springer-Verlag, 1997), pp. 299–301, 520–528.

The four-winged fruit fly

Thomas Hunt Morgan, Calvin B. Bridges, and A. H. Sturtevant, *The Genetics of Drosophila,* Reprint Edition (New York: Garland Publishing, 1988; originally S'Gravenhage, Netherlands: M. Nijhoff, 1925), p. 79; F. H. C. Crick and P. A. Lawrence, "Compartments and Polyclones in Insect Development," *Science* 189 (1975), pp. 340–347; E. B. Lewis, "A gene complex controlling segmentation in *Drosophila*," *Nature* 276 (1978), pp. 565–570; E. B. Lewis, "Control of Body Segment Differentation in *Drosophila* by the Bithorax Gene Complex," pp. 269–288 in Max M. Burger, and Rudolf Weber (editors), *Embryonic Development, Part A: Genetic Aspects* (New York, Alan R. Liss, 1982); E. B. Lewis, "Regulation of the Genes of the Bithorax Complex in *Drosophila*," *Cold Spring Harbor Symposia on Quantitative Biology* 50 (1985), pp. 155–164; Jordi Casanova, Ernesto Sánchez-Herrero, and Ginés Morata, "Prothoracic Transformation and Functional Structure of the *Ultrabithorax* Gene of Drosophila," *Cell* 42 (1985), pp. 663–669; Philip A. Beachy, "A molecular view of the *Ultrabithorax* homeotic gene of *Drosophila*," *Trends in Genetics* 6 (1990), pp. 46–51.

Figure 9–2 data are from Mark Peifer and Welcome Bender, "The anterobithorax and bithorax mutations of the bithorax complex," *EMBO Journal* 5 (1986), pp. 2293–2303; E. B. Lewis, "Genes and Developmental Pathways," *American Zoologist* 3 (1963), pp. 33–56.

Four-winged fruit flies and evolution

The textbook quotation is from Peter H. Raven and George B. Johnson, *Biology*, Fifth Edition (Boston: WCB/McGraw-Hill, 1999), p. 334. See also William K. Purves, Gordon H. Orians, H. Craig Heller, and David Sadava, *Life: The Science of Biology*, Fifth Edition (Sunderland, MA: Sinauer Associates, 1998), pp. 508–509; Douglas Futuyma, *Evolutionary Biology*, Third Edition (Sunderland, MA: Sinauer Associates, 1998), pp. 48–49.

On the absence of flight muscles in the second pair of wings, see H. H. El Shatoury, "Developmental Interactions in the Development of the Imaginal Muscles of *Drosophila*," *Journal of Embryology and Experimental Morphology* 4 (1956), pp. 228–239; Alberto Ferrus and Douglas R. Kankel, "Cell Lineage Relationships in *Drosophila melanogaster:* The Relationships of Cuticular to Internal Tissues," *Developmental Biology* 85 (1981), pp. 485–504; M. David Egger, Suzan Harris, Bonnie Peng, Anne M. Schneiderman, and Robert J. Wyman, "Morphometric Analysis of Thoracic Muscles in Wildtype and in Bithorax *Drosophila*," *The Anatomical Record* 226 (1990), pp. 373–382; J. Fernandes, S. E. Celniker, E. B. Lewis, and K. VijayRaghavan, "Muscle development in the four-winged *Drosophila* and the role of the *Ultrabithorax* gene," *Current Biology* 4 (1994), pp. 957–964; Sudipto Roy, L. S. Shashidhara, and K. VijayRaghavan, "Muscles in the *Drosophila* second thoracic segment are patterned independently of autonomous homeotic gene function," *Current Biology* 7 (1997), pp. 222–227.

For Mayr's critique of macromutations, see Ernst Mayr, *Populations, Species and Evolution*, an abridgement of his 1963 book, *Animal Species and Evolution* (Cambridge, MA: Harvard University Press, 1970), pp. 251–253. Mayr was criticizing the view of Berkeley geneticist Richard Goldschmidt that major mutations—producing what Goldschmidt called "hopeful monsters"—might overcome the inability of small mutations to account for evolution. See Peter J. Bowler, *Evolution: The History of an Idea*, Revised Edition (Berkeley, CA: University of California Press, 1989), pp. 339–340.

Evolution in reverse?

National Academy of Sciences, *Teaching About Evolution and the Nature of Science* (Washington, DC: National Academy Press, 1998), Chapter 5, p. 2. See also National Academy of Sciences, *Science and Creationism: A View from the National Academy of Sciences*, Second Edition (Washington, DC: National Academy of Sciences Press, 1999), Appendix, p. 1.

On the complex network of interactions controlled by *Ultrabithorax*, see Scott D. Weatherbee, Georg Halder, Jaeseob Kim, Angela Hudson, and Sean Carroll, "*Ultrabithorax* regulates genes at several levels of the wing-patterning hierarchy to shape the development of the *Drosophila* haltere," *Genes & Development* 12 (1998), pp. 1474–1482.

Are DNA mutations the raw materials for evolution?

Textbook quotations are from Cecie Starr and Ralph Taggart, *Biology: The Unity and Diversity of Life*, Eighth Edition (Belmont, CA: Wadsworth Publishing Company, 1998), p. 283; Burton S. Guttman, *Biology* (Boston: WCB/McGraw-Hill, 1999), p. 470.

On saturation mutagenesis in fruit flies, see Christiane Nüsslein-Volhard and Eric Wieschaus, "Mutations affecting segment number and polarity in *Drosophila*," *Nature* 287 (1980), pp. 795–801; Daniel St. Johnston and Christiane Nüsslein-Volhard, "The Origin of Pattern and Polarity in the *Drosophila* Embryo," *Cell* 68 (1992), pp. 201–219. On saturation mutagenesis in zebrafish, see Peter Aldhous, "'Saturation screen' lets zebrafish show their stripes," *Nature* 404 (2000), p. 910; Gretchen Vogel, "Zebrafish Earns Its Stripes in Genetic Screens," *Science* 288 (2000), pp. 1160–1161.

Beyond the gene

For recent publications questioning whether genes control development, see B. C. Goodwin, "What are the Causes of Morphogenesis?" *BioEssays* 3 (1985), pp. 32–36; J. M. Barry, "Informational DNA: a useful concept?" *Trends in Biochemical Sciences* 11 (1986), pp. 317–318; Michael Locke, "Is there somatic inheritance of intracellular patterns?" *Journal of Cell Science* 96 (1990), pp. 563–567; H. F. Nijhout, "Metaphors and the Role of Genes in Development," *BioEssays* 12 (1990), pp. 441–446; Jonathan Wells, "The History and Limits of Genetic Engineering," *International Journal on the Unity of the Sciences* 5 (1992), pp. 137–150; Brian C. Goodwin, *How the Leopard Changed Its Spots* (New York: Charles Scribner's Sons, 1994).

On the rise of the neo-Darwinian monopoly in genetics see Jan Sapp, *Beyond the Gene: Cytoplasmic Inheritance and the Struggle for Authority in Genetics* (Oxford: Oxford University Press, 1987); quotations are from pp. 59, 81, 85.

The Conference on "Genes and Development" was sponsored by the Institut für Ethik und Geschichte der Medizin in Basel,

Switzerland, March 19–20, 1999. I am not mentioning the name of the German participant who told me her story, for the same reason I am withholding the name of the Chinese paleontologist mentioned at the end of the Tree of Life chapter—to protect her from Darwinian heresy-hunters.

Chapter 10: Fossil Horses and Directed Evolution

Figure 10–1 is from William D. Matthew, "The Evolution of the Horse," Supplement to *American Museum of Natural History Journal* 3 (January 1903), Guide Leaflet No. 9, following p. 8.

Fossil horses and orthogenesis

On orthogenesis, see Ernst Mayr, *The Growth of Biological Thought* (Cambridge, MA: Harvard University Press, 1982), pp. 528–531; Peter J. Bowler, Evolution: *The History of an Idea*, Revised Edition (Berkeley, CA: University of California Press, 1989), pp. 268–270; Robert C. Richardson and Thomas C. Kane, "Orthogenesis and Evolution in the 19th Century: The Idea of Progress in American Neo-Lamarckism," pp. 149–167 in Matthew H. Nitecki (editor), *Evolutionary Progress* (Chicago: The University of Chicago Press, 1988).

The Schindewolf quotations are from Otto H. Schindewolf, *Basic Questions in Paleontology* (originally *Grundfragen der Paläontologie*, 1950), translated by Judith Schaefer (Chicago: The University of Chicago Press, 1993), p. 270, 273.

The Simpson quotation is from George Gaylord Simpson, *The Meaning of Evolution* (New Haven, CT: Yale University Press, 1949), p. 159.

Revising the picture of horse evolution

The single most complete resource for information on horse evolution is Bruce J. McFadden, *Fossil Horses: Systematics, Paleobiology, and Evolution of the Family Equidae* (Cambridge: Cambridge University Press, 1992). For an earlier summary, see Bruce J. McFadden, "Horses, the Fossil Record, and Evolution," *Evolutionary Biology* 22 (1988), pp. 131–158. See also Robert J. G. Savage, *Mammal Evolution: An Illustrated Guide* (New York & Oxford: Facts on File and The British Museum [Natural History]: 1986), pp. 200–205.

The Simpson quotation is from George Gaylord Simpson, *Tempo and Mode in Evolution* (New York, Columbia University Press, 1944), p. 163.

Figure 10–2 data are from Bruce J. McFadden, *Fossil Horses,* pp. 99, 194.

On *Miohippus* preceding *Mesohippus*, see Donald R. Prothero and Neil Shubin, "The Evolution of Oligocene Horses," pp. 142–175 in Donald R. Prothero and Robert M. Schoch (editors), *The Evolution of Perissodactyls* (New York: Oxford University Press, 1989), p. 151; Bruce J. McFadden, *Fossil Horses,* p. 176.

What does the evidence really show?

For Simpson's views, see George Gaylord Simpson, *The Meaning of Evolution* (New Haven, CT: Yale University Press, 1949), p. 159; Simpson, *Horses* (New York: Oxford University Press, 1951); Simpson, *The Major Features of Evolution* (New York: Simon & Schuster, 1953), pp. 260–265.

Undirected evolution from Darwin to Dawkins

Darwin's quotations are from Francis Darwin (editor), *The Life and Letters of Charles Darwin* (New York: D. Appleton, 1887), Vol. I, pp. 278–279; Vol. II, pp. 105–106. See also Francis Darwin and

A. C. Seward (editors), *More Letter of Charles Darwin* (New York: D. Appleton, 1903), Vol. I, pp. 191–192, 321, 395; Francis Darwin (editor), *The Life and Letters of Charles Darwin*, Vol. II, pp. 97–98. 146, 169–170, 247.

On Darwin's opposition to directed evolution, see Neal C. Gillespie, *Charles Darwin and the Problem of Creation* (Chicago: The University of Chicago Press, 1979); Jonathan Wells, "Charles Darwin on the Teleology of Evolution," *International Journal on the Unity of the Sciences* 4 (1991), pp. 133–156.

The Simpson quotations are from Simpson, *The Meaning of Evolution*, pp. 132, 345. The Monod quotation is from Horace Freeland Judson, *The Eighth Day Of Creation* (New York: Simon & Schuster, 1979), p. 217.

The blind watchmaker

William Paley, *Natural Theology*, Reprint of 1802 edition (Houston, TX: St. Thomas Press, 1972), p. 1; Richard Dawkins, *The Blind Watchmaker* (New York: W. W. Norton, 1986), pp. x, 1, 5, 6, 287.

Teaching materialistic philosophy in the guise of science

Textbook quotations are from Kenneth R. Miller and Joseph S. Levine, *Biology*, Fifth Edition (Upper Saddle River, NJ: Prentice-Hall, 2000), p. 658; William K. Purves, Gordon H. Orians, H. Craig Heller, and David Sadava, *Life: The Science of Biology*, Fifth Edition (Sunderland, MA: Sinauer Associates, 1998), p. 10; Neil A. Campbell, Jane B. Reece, and Lawrence G. Mitchell, *Biology*, Fifth Edition (Menlo Park, CA: The Benjamin/Cummings Publishing Company, 1999), pp. 412–413; Peter H. Raven and George B. Johnson, *Biology*, Fifth Edition (Boston: WCB/ McGraw-Hill, 1999), p. 15; Douglas J. Futuyma, *Evolutionary Biology*, Third Edition (Sunderland, MA: Sinauer Associates, 1998), pp. 8, 5.

Chapter 11: From Ape to Human: The Ultimate Icon

Darwin, *The Origin of Species*, Conclusion, p. 373; Stephen Jay Gould, *Wonderful Life* (New York: W. W. Norton, 1989), p. 28.

Are we (just) animals?

Darwin, *The Descent of Man*, pp. 395, 445, 446, 456, 494, 471–472, 469, 470. The Dawkins quotation is from Richard Dawkins, "Darwinism and human purpose," pp. 137–143, in John R. Durant (editor), *Human Origins* (Oxford: Clarendon Press, 1989), pp. 137–138. Thomas Aquinas lists many senses and emotions that are "common to men and other animals" in his *Summa Theologiae*, First Part (Treatise on Man) and First Part of the Second Part (Treatise on the Divine Government).

Finding evidence to fit the theory

Thomas Henry Huxley, *Evidence as to Man's Place in Nature*, reprint of 1863 edition (New York: D. Appleton, 1886); quotations are from pp. 125 & 132 of the 1886 edition.

On Neanderthals, see Marcellin Boule and Henri V. Vallois, *Fossil Men: A Textbook of Human Palaeontology* (London: Thames and Hudson, 1957), originally *Les Hommes Fossiles* (1923); a comparison of Boule's stooped reconstruction of a Neanderthal skeleton and a modern human is on p. 253. See also Niles Eldredge and Ian Tattersall, *The Myths of Human Evolution* (New York: Columbia University Press, 1982), p. 76.

Excluding Neanderthal from the human lineage helped to prepare the way for Piltdown. See Michael Hammond, "A Framework of Plausibility for an Anthropological Forgery: The Piltdown Case," *Anthropology* 3 (1979), pp. 47–58; Lewin, *Bones of Contention*, pp. 63–70; Ian Tattersall, *The Fossil Trail: How We Know What We*

Think We Know About Human Evolution (New York: Oxford University Press, 1995), pp. 20–24, 36–39.

The Piltdown fraud

The original published report of Piltdown was C. Dawson and A. S. Woodward, "On the discovery of a Palaeolithic human skull and mandible in a flint-bearing gravel overlying the Wealden (Hastings Beds) at Piltdown, Fletching (Sussex)," *Quarterly Journal of the Geological Society of London* 69 (1913), pp. 117–151.

On the exposure of the Piltdown fraud, see J. S. Weiner, F. P. Oakley, and W. E. Le Gros Clark, "The Solution of the Piltdown Problem," *Bulletin of the British Museum (Natural History), Geology* 2 (1953), pp. 139–146; and J. S. Weiner, et al., "Further Contributions to the Solution of the Piltdown Problem," *Bulletin of the British Museum (Natural History), Geology* 2 (1953), pp. 225–287; J. S. Weiner and K. P. Oakley, "The Piltdown Fraud: Available Evidence Reviewed," *American Journal of Physical Anthropology* 12 (1954), pp. 1–7; J. S. Weiner, *The Piltdown Forgery* (London: Oxford University Press, 1955). Some theories as to who the perpetrators might have been are in Ronald Millar, *The Piltdown Men* (London: Victor Gollancz, 1972); Stephen Jay Gould, "The Piltdown Conspiracy," *Natural History* 89 (August 1980), pp. 8–28; John Hathaway Winslow and Alfred Meyer, "The Perpetrator at Piltdown," *Science 83* 4 (September 1983), pp. 33–43.

Quotations are from Lewin, *Bones of Contention*, pp. 70, 73; Jane Maienschein, "The One and the Many: Epistemological Reflections on the Modern Human Origins Debate," pp. 413–422 in G. A. Clark and C. M. Willermet (editors), *Conceptual Issues in Modern Human Origins Research* (New York: Aldine de Gruyter, 1997), p. 415; John Napier, *The Roots of Mankind* (Washington, DC: Smithsonian Institution Press, 1970), p. 139; Eldredge and Tattersall, *The Myths of Human Evolution*, pp. 126–127.

How much can the fossils show us?

On the variable appearance of skull 1470, see Lewin, *Bones of Contention*, p. 160; see also Tattersall, *The Fossil Trail*, p. 133. The drawings of *Homo habilis* by four different artists are in "Behind the Scenes," *National Geographic* 197 (March, 2000), p. 140. The drawings are actually on an unnumbered page, buried among the advertisements at the end of the issue; the page number cited here was obtained by extrapolating from the last numbered page.

On the difficulty of reconstructing evolutionary history, see Constance Holden, "The Politics of Paleoanthropology," *Science* 213 (1981), pp. 737–740. The Gee quotations are from Henry Gee, *In Search of Deep Time: Beyond the Fossil Record to a New History of Life* (New York: The Free Press, 1999), pp. 113, 23, 32, 202, 32, 116–117.

Paleoanthropology: science or myth?

On Durant's comments, see Lewin, *Bones of Contention*, p. 312; John R. Durant, "The myth of human evolution," *New Universities Quarterly* 35 (1981), pp. 425–438. For the report on Matt Cartmill's remarks, April 13, 1984, see Lewin, *Bones of Contention*, p. 302.

Misia Landau, *Narratives of Human Evolution* (New Haven, CT: Yale University Press, 1991), pp. ix–x, 148.

For the Tattersall and Clark quotations, see Ian Tattersall, "Paleoanthropology and Preconception," pp. 47–54 in W. Eric Meikle, F. Clark Howell and Nina G. Jablonski (editors), *Contemporary Issues in Human Evolution*, Memoir 21 (San Francisco, CA: California Academy of Sciences, 1996), p. 53; Geoffrey A. Clark, "Through a Glass Darkly: Conceptual Issues in Modern Human Origins Research," pp. 60–76 in G. A. Clark and C. M. Willermet (editors), *Conceptual Issues in Modern Human Origins Research* (New York: Aldine de Gruyter, 1997), p. 76.

What do we know about human origins?

For good drawings of many of the major fossil skulls now known, see Ian Tattersall, *The Fossil Trail: How We Know What We Think We Know About Human Evolution* (New York: Oxford University Press, 1995). For a collection of essays by parties to recent controversies over human origins, see Russell L. Ciochon and John G. Fleagle, *The Human Evolution Source Book* (Englewood Cliffs, NJ: Prentice Hall, 1993). See also other essays in the two collections cited below.

The quotation about 150 different views on Neanderthals is from James Shreeve, *The Neandertal Enigma* (New York: William Morrow, 1995), p. 252.

F. Clark Howell, "Thoughts on the Study and Interpretation of the Human Fossil Record," pp. 1–39 in W. Eric Meikle, F. Clark Howell, and Nina G. Jablonski (editors), *Contemporary Issues in Human Evolution*, pp. 3, 31.

Geoffrey A. Clark, "Through a Glass Darkly: Conceptual Issues in Modern Human Origins Research," pp. 60–76 in G. A. Clark and C. M. Willermet (editors), *Conceptual Issues in Modern Human Origins Research*, p. 60–62

Concepts masquerading as neutral descriptions of nature

Stephen Jay Gould, *Wonderful Life* (New York, W. W. Norton, 1989), pp. 27–52. The Ruse quotations are from Michael Ruse, "How evolution became a religion," *National Post* (May 13, 2000), www.nationalpost.com/artslife.asp?f=000513/288424.html. The textbook interview with Gould is in Peter H. Raven and George B. Johnson, *Biology*, Fifth Edition (Boston: WCB/ McGraw-Hill, 1999). p. 14.

Chapter 12: Science or Myth?

The Mayr quotation is from Ernst Mayr, "Darwin's Influence on Modern Thought," *Scientific American* 283 (July 2000), pp. 79–83.

The "F" word

The Gould quotation is from Stephen Jay Gould, "Abscheulich! Atrocious!" *Natural History* (March 2000), pp. 42–49.

The Ritter quotation is from Carla Yu, "Moth-eaten Darwinism: A disproven textbook case of natural selection refuses to die," *Alberta Report Newsmagazine*, Vol. 26, No. 15 (April 5, 1999), pp. 38–39. The textbook in question is Bob Ritter, Richard F. Coombs, R. Bruce Drysdale, Grant A. Gardner, and Dave T. Lunn, *Biology* (Scarborough, ONT: Nelson Canada, 1993), which deals with peppered moths on pp. 109–110.

The impenetrable disclaimer about "conceptual integumentary structures" on *Bambiraptor* is from David A. Burnham, Kraig L. Derstler, Philip J. Currie, Robert T. Bakker, Zhonghe Zhou, and John H. Ostrom, "Remarkable New Birdlike Dinosaur (Theropoda: Maniraptora) from the Upper Cretaceous of Montana," *The University of Kansas Paleontological Contributions*, New Series, No. 13 (March 15, 2000), p. 8.

Scientific misconduct and stock fraud

National Academy of Sciences, *Teaching About Evolution and the Nature of Science* (Washington, DC: National Academy Press, 1998), Chapter 2, p. 10; National Academy of Sciences, *Science and Creationism: A View from the National Academy of Sciences*, Second Edition (Washington, DC: National Academy of Sciences Press, 1999), Chapter on "Evidence Supporting Biological Evolution," p. 2.

Phillip E. Johnson, "The Church of Darwin," *The Wall Street Journal* (August 16, 1999), p. A14. On applying securities law to scientific misconduct, see Securities Exchange Act of 1934,

17 C.F.R. 240.10b–5; Louis M. Guenin, "Expressing a consensus on candour," *Nature* 402 (1999), pp. 577–578.

Darwinian censorship

On efforts by the Baylor University faculty to silence criticism of Darwinism, see Mark Wingfield, "Baylor faculty, administration clash over center for creation study," *Associated Baptist News* (May 12, 2000); John Drake, "Sloan nixes decision to dissolve Polanyi," *Baylor Lariat* (Thursday, April 20, 2000). See also Ron Nissimov, "Baylor professors concerned center is front for promoting creationism," *Houston Chronicle*, July 3, 2000, p. 1 (www.chron.com/cs/CDA/story.hts/page1). For information on the Michael Polanyi Center check their web site, www.baylor.edu/~polanyi.

On efforts by the Melvindale, Michigan, school board to place books critical of Darwinism in the high school library, see Jonathan Wells, "Local book battle concerns academic liberty," *The Detroit News* (March 14, 1999), p. 7B. The NCSE quotations are from Molleen Matsumura, "Facing Challenges to Evolution Education," http://www.natcenscied.org/tenchal.html.

The Roger DeHart story is summarized in a newspaper article by Theresa Goffredo, "School officials throw extra science materials out of class," *The* [Burlington, WA] *Skagit Valley Herald* (May 28, 2000), p. A1. Internet version: http://www.skagitvalleyherald.com/daily/00/may/28/a1dehart.html.

The standards actually adopted by Kansas are at the Board of Education's internet web site: http://www.ksbe.state.ks.us/outcomes/science_12799.html. On what really happened in Kansas, see Jonathan Wells, "Ridiculing Kansas school board easy, but it's not good journalism," *The* [Mitchell, SD] *Daily Republic* (October 14, 1999). The text of this op-ed is reproduced here, by permission:

Wizard of Oz jokes are in vogue as the news media scramble to ridicule Kansas for downplaying, eliminating, or even banning evolution in its public schools. But the people who are writing such stuff apparently haven't read the Kansas Science Education Standards. The truth is that the August 11 School Board decision actually increased public school emphasis on evolution.

The old science standards, in effect since 1995, devoted about 70 words to biological evolution. Standards proposed to the Board earlier this year by a 27-member Science Education Standards Writing Committee would have increased this to about 640 words. The standards actually adopted by the Board on August 11 include about 390 words on the subject. So the Kansas State School Board, asked to approve a ninefold increase in the standards for evolution, approved a fivefold increase instead.

Of course, word counts don't tell the whole story. But the 390 words approved by the Board include many of the provisions recommended by the Committee. For example, the Board adopted verbatim the Committee's summary of Darwin's theory: "Natural selection includes the following concepts: 1) Heritable variation exists in every species; 2) some heritable traits are more advantageous to reproduction and/or survival than are others; 3) there is a finite supply of resources available for life; not all progeny survive; 4) individuals with advantageous traits generally survive; 5) the advantageous traits increase in the population through time." It would be difficult to find a better summary of Darwin's theory of natural selection; Kansas students will now be tested on it.

The Board also required students to understand that "microevolution… favors beneficial genetic variations and contributes to biological diversity," and listed finch beak changes as an example. The Board declined, however, to adopt the Committee's proposal requiring students to understand that microevolution leads to macroevolution—the origin of new structures and new groups of organisms. The Board's reluctance is understandable, since even some biologists doubt that changes in finch beaks can explain the origin of finches in the first place.

There were some other recommendations the Board did not follow, as well. For example, the Committee would have required students to understand: "The common ancestry of living things allows them to be classified into a hierarchy of groups." This requirement would no doubt have come as a surprise to 18th century creationist Carolus Linnaeus, who had no need of common ancestry when he devised the hierarchical system of classification still used by modern biologists.

Even more interesting than the details, however, was the Committee's bid to inject evolution into the very heart of science. According to the 1995 standards, science embodies four general themes: Energy/Matter, Patterns of Change, Systems and Interactions, and Stability and Models. Furthermore, it is the nature of science to "provide a means for producing knowledge," using processes such as "observing, classifying, questioning, inferring,… [and] collecting and recording data." The Science Education Standards Writing Committee proposed to add a fifth general theme, "patterns of cumulative change," an example of which is "the biological theory of evolution."

As a biologist myself, I find this strange. Why list a specific theory such as biological evolution among general themes such as "systems and interactions," or basic processes such as "collecting and recording data"? That's like inserting a specific law into a constitution designed to establish a framework for lawmaking.

Why did the 1995 standards have to be changed at all? The Committee's proposal was a product of recent nationwide efforts by people who believe that Darwinian evolution is indispensable to biological science. A rallying cry for these efforts is Theodosius Dobzhansky's famous maxim, "Nothing in biology makes sense except in the light of evolution." But Dobzhansky was mistaken. There are entire areas of biology that have no need for evolutionary theory, and there is evidence that the most sweeping claims of Darwinism are wrong. More importantly, there can be no such thing as an indispensable theory in science. A true scientist would say that nothing in biology makes sense except in the light of evidence.

The standards adopted by the Kansas State School Board are far from perfect. Biology education would have been better served if students had been required to understand macroevolutionary theory, though they should also be taught the scientific evidence against it. Under the circumstances, however, the Board may have done the best it could. Faced with national pressure to include Darwin's theory in its description of the very nature of science, the Board courageously resisted, stocking the shelves with more evolution but refusing to hand over the store.

News commentators who ridicule Kansas for downplaying, eliminating, or even banning evolution from its schools not only misrepresent the truth, but they also miss the real story. Why do Darwinists go ballistic at the thought of high school students questioning their theory? Why do biology textbooks continue to cite evidence for evolution that was long ago discredited? How many qualified scientists have lost their teaching jobs or their research funding just because they dared to criticize Darwinism? How many millions of your tax dollars will be spent this year by Darwinists trying to find evidence for a theory they claim is already proven beyond a reasonable doubt? There's enough here to keep a team of investigative journalists busy for months.

Years ago, when asked why the media were spending so much time covering the O. J. Simpson trial, a news commentator said, "It's easy work." Ridiculing Kansas is easy work, too. But it's not good journalism.

On denying Kansas schoolchildren admission to Darwinist-controlled colleges, see Herbert Lin, "Kansas Evolution Ruling," *Science* 285 (1999), p. 1849; John Rennie, "A Total Eclipse of Reason," *Scientific American* 281 (October 1999), p. 124; and John Rennie, "Fan Mail from the Fringe," *Scientific American* (February 2000), p. 4.

Bruce Alberts wrote the preface for the booklet, *Science and Creationism: A View from the National Academy of Sciences*, Second Edition (Washington, DC: National Academy of Sciences Press, 1999); Alberts is also the first author on *Molecular Biology of the Cell,* Third Edition (New York: Garland Publishing, 1994), which features the Miller-Urey experiment (p. 4) and Haeckel's embryos (p. 33).

It's your money

The information on federal science funding for 2000 is from David Malakoff, "Balancing the Science Budget," *Science* 287 (February 11, 2000), pp. 952–955. The Federation of American Societies for Experimental Biology is lobbying Congress to double NIH funding in 2001; see *FASEB News* 33 (April 2000), p. 1. Futuyma's "rumor" that the NSF tells grant applicants to omit the word "evolution" is from Douglas Futuyma, "Evolution as Fact and Theory," *Bios* 56 (1985), pp. 3–13.

What can you do about it?

The Guenin quotation is from Louis M. Guenin, "Expressing a consensus on candour," *Nature* 402 (1999), pp. 577–578. Information on funding sources for the National Academy is from http://www.nationalacademies.org/about/faq.

U.S. Representative Mark Souder's remarks are in the Congressional Record for June 14, 2000, p. H4480.

Nothing in biology makes sense except in the light of WHAT?

Theodosius Dobzhansky's maxim is from "Nothing in Biology Makes Sense Except in the Light of Evolution," *The American Biology Teacher* 35 (1973), pp. 125–129. Peter Grant's comment is from "What Does It Mean to Be A Naturalist at the End of the Twentieth Century?" *The American Naturalist* 155 (2000), pp. 1–12.

Index